D0292100

Immobilized Enzymes

Immobilized Enzymes

An Introduction and Applications in Biotechnology

Michael D. Trevan
Senior Lecturer in Biochemistry
The Hatfield Polytechnic

JOHN WILEY & SONS
Chichester · New York · Brisbane · Toronto

British Library Cataloguing in Publication Data:

Trevan, Michael D.
Immobilized enzymes.
1. Immobilized enzymes
I. Title
547'.758 QP601 80-40502

ISBN 0 471 27826 2

Phototypeset by Dobbie Typesetting Service, Plymouth, Devon, England and printed in the United States of America.

To Marilyn and all others
who demand that science should be intelligible,
I hope this helps

Contents

Abbreviations and Symbols

A	Surface area
D	Diffusion coefficient
d	Pathlength of diffusion
E_T	Total enzyme concentration
H_e^+	Hydrogen ion concentration in bulk phase
H_i^+	Hydrogen ion concentration in enzyme phase (microenvironment)
h_s	Substrate transport coefficient
J_{H^+}	Rate of flux of hydrogen ions
K'	1st order rate constant
K_v	Half-saturation constant (substrate concentration at $V_s/2$)
p	Partition coefficient
P	Product concentration
pH_e	pH of the bulk phase
pH_i	pH of the enzyme phase (microenvironment)
S	Substrate concentration
S_o	Substrate concentration in bulk phase
S_i	Substrate concentration in enzyme phase (microenvironment)
V	Reaction velocity
V'_{max}	Saturation velocity at pH optimum
V_s	Saturation velocity for an immobilized enzyme
V(diff)	Diffusion velocity of substrate
Δ	Ionic partition coefficient

Preface

The immobilization of enzymes by fixing them in some way on to an inert and usually insoluble polymer matrix, is an area of research that is currently generating much interest among biochemists, organic and physical chemists, microbiologists, biomathematicians, biophysicists, and chemical engineers. It is probably because of this multidisciplinary approach that most of the literature on the subject is written by and for the active research worker and is largely inaccessible to the 'non-expert'. This book is an attempt to fill that gap and provide an introductory text to this fascinating subject; it is not intended as a reference book, sufficient of these exist. Unnecessary and confusing detail is thus omitted and explanation of many of the effects of immobilization is kept as unmathematical as is humanly possible. Understanding, therefore, is based on the reader gaining a thorough intuitive grasp of the concepts involved.

Research on immobilized enzymes has been carried out for the past two decades from two principal standpoints. First it was apparent that there was vast potential for using immobilized enzymes as recoverable, stable, and specific industrial catalysts. A great deal of research has thus gone into the immobilization of enzymes of commercial importance and the associated problems of reactor design and scale-up procedures. However, despite the quantity and quality of the research in this field the potential has as yet been little realized.

Second, immobilized enzymes have been studied with a view to discover some of the effects a heterogeneous environment may have on an enzyme-catalysed reaction. Study of highly purified enzymes in dilute solution under Michaelis–Menten conditions reveals a great deal about the way in which enzymes will behave under such conditions. Unfortunately for the enzymologist, most intracellular enzymes are neither working under Michaelis–Menten conditions nor are they in dilute solution, but in a complex heterogeneous environment. Thus for a better understanding of the behaviour and control of enzymes functioning *in vivo*, many attempts have been made to study the effect of immobilization and a defined heterogeneous environment on the enzyme and then relate this to the *in vivo* situation.

This book is divided into five chapters. Chapter 1 covers briefly some of the methods by which enzymes may be immobilized and the factors that lead to the choice of a particular method. Chapter 5 details some easy methods for preparing immobilized enzymes. Chapter 2 relates some of the effects that immobilization may have on the kinetic parameters of an enzyme-catalysed reaction. Chapter 3 explores some of the possibilities and problems in the commercial and medical applications of immobilized enzymes. In Chapter 4 an attempt will be made to illustrate the way in which immobilized enzyme systems can be made to mimic biological systems. The information obtained from such model systems will be discussed with particular reference to its relevance to our understanding of the cell and its biochemistry.

I would like to acknowledge the help and cooperation of all my colleagues during the period of the preparation of this book. In particular my thanks to Elaine Johnson and Vyvyan Ellen for their help in the preparation of this manuscript. I would also like to thank my wife, Marilyn, for her understanding, enthusiasm, encouragement and endless cups of late night coffee without which I doubt that this project would have come to fruition.

Chapter 1

Techniques of Immobilization

I. IMMOBILIZATION — THE PRINCIPLES AND THE PROCESS

Enzyme immobilization may be defined as the imprisonment of an enzyme molecule in a distinct phase that allows exchange with, but is separated from, the bulk phase in which substrate effector or inhibitor molecules are dispersed and monitored. The enzyme phase (see Figure 1) is usually insoluble in water and is often a high molecular weight, hydrophilic polymer (e.g. cellulose). The imprisonment of the enzyme may be achieved by various means. The enzyme can be covalently bonded to, adsorbed on to, or physically entrapped within the enzyme phase.

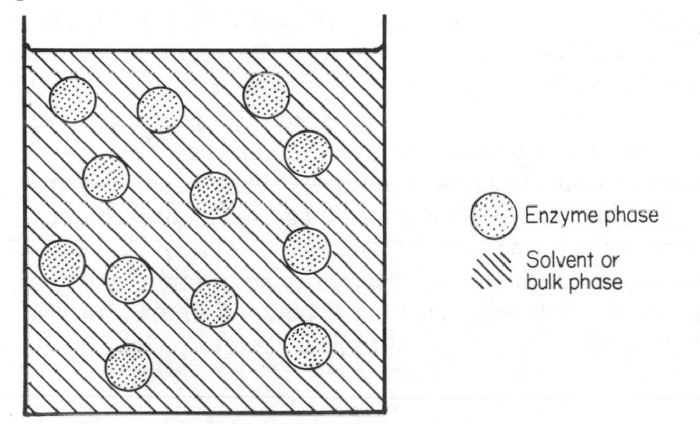

Figure 1. Definition of enzyme and bulk phases in immobilized enzyme system

The use of the term 'immobilization' can lead to confusion, for it implies, incorrectly, that the enzyme molecule can never move about within its distinct phase. Immobilization is, however, a more apt term for the process than the once much used term 'insolubilization' as water-soluble immobile enzymes can be prepared. The reader must, however, beware as both terms are used interchangeably in the literature.

The nature of the immobilized enzyme preparation will obviously depend upon the character of the enzyme phase. For example, the most common form of immobilized enzyme is that in which the enzyme molecule has been covalently bonded on to an insoluble polymer such as cellulose or polyacrylamide.

The polymer, however, may be in the form of a particulate powder or in sheet form as a membrane. The enzyme may even be covalently bonded on to itself (or another inert protein), forming an insoluble, but active, polymeric enzyme. Yet another approach is to attach the enzyme to a polymer using electrostatic or other non-covalent bonding mechanisms. Alternatively the enzyme need not be bound to the polymer, but may be trapped within it, the polymer forming a net-like matrix around the enzyme, the pores in the net being too small to allow the escape of the enzyme, but large enough to permit the entry of low molecular weight substrates. A variation of this last technique lies in the use of phospholipid bilayers as the enzyme phase. In this case the enzyme may either be in aqueous solution surrounded by a phospholipid barrier (see liposomes p.64) or actually be 'dissolved' in the hydrophobic portion of the bilayer (see p.117). Figure 2 depicts diagrammatically these various forms of immobilized enzyme.

Immobilization often causes a dramatic change in the apparent measured parameters of the enzyme-catalysed reaction. For example, the maximum velocity of reaction, Michaelis–Menten constant, temperature optimum, pH optimum, effect of inhibitors may all be changed when an enzyme is immobilized; the degree and nature of this change will depend not only on the immobilization method used, but also on the enzyme reaction. These effects are of great interest to the biochemist who wishes to use immobilized enzymes as model systems of enzyme action in the cellular environment, but they may be a constant source of irritation to the industrialist who wishes to use immobilized enzymes as efficient, specific, recoverable catalysts. The industrial uses of immobilized enzymes, the manner in which immobilization may affect an enzyme reaction, and the way immobilized enzymes can be used as models of biological systems are all discussed in later chapters. The rest of this chapter is concerned with an account of some of the methods that have been used to prepare immobilized enzymes and, finally, the factors affecting the choice of an appropriate method. Chapter 5 outlines several methods of immobilization in sufficient detail to provide easy recipes for the uninitiated to try.

II. COVALENT BONDING TO ACTIVATED POLYMERS

Immobilization by covalent bonding to activated polymers is probably the most extensively used method, for although often tedious it provides an immobile enzyme that is firmly bound to its polymeric support. The range of polymers and chemical coupling procedures used is enormous and so the descriptions here cover only the most important methods developed over the years.

1. The Emergence of the Technique

The history of enzyme immobilization goes back to the late 1940s. In 1949

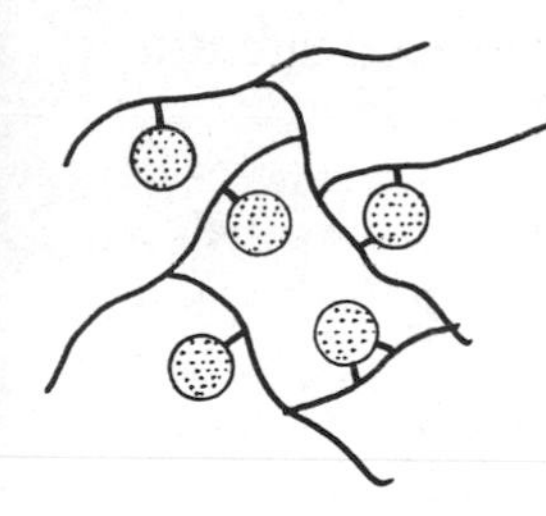

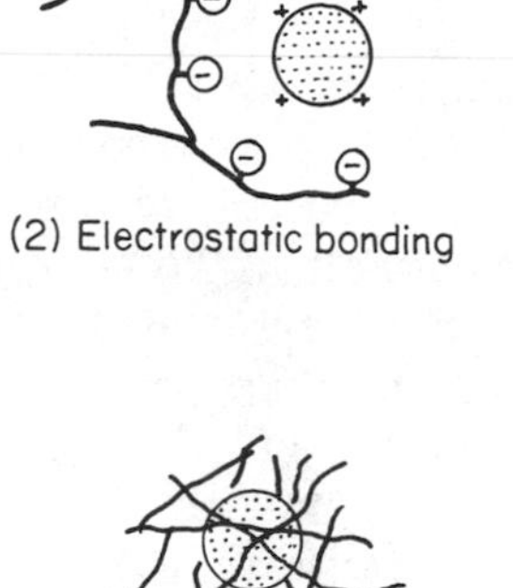

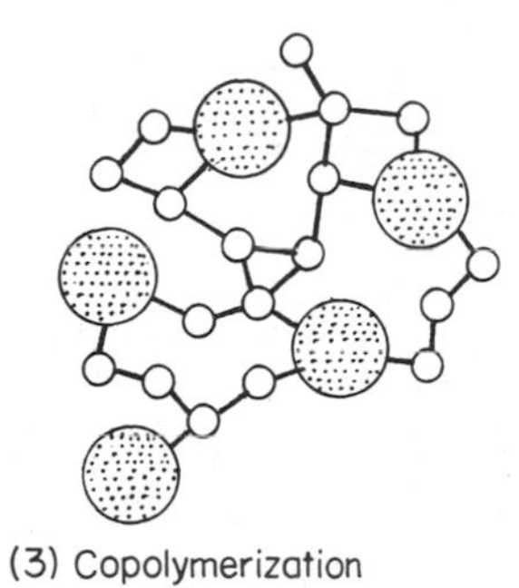

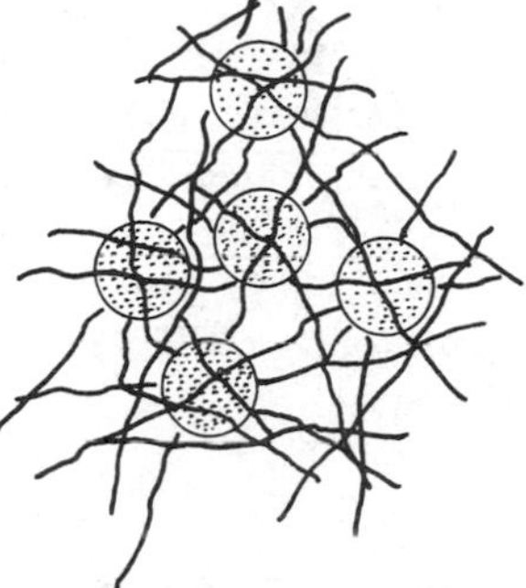

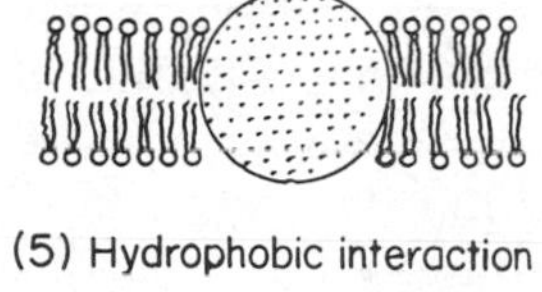

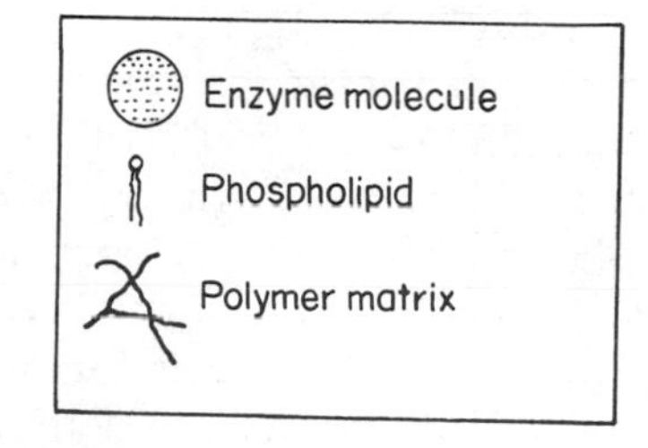

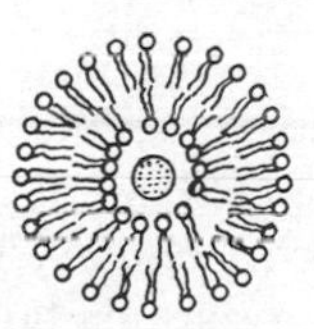

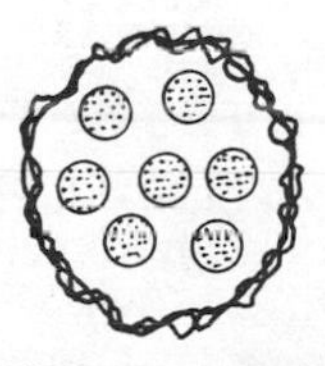

Figure 2. Possible modes of enzyme immobilization

Michael and Ewers used the azide derivative of carboxymethylcellulose to immobilize a variety of proteins. However, it was not until Grubhofer and Schleith, who in 1954 used a diazo derivative of poly-*p*-aminostyrene to immobilize pepsin, amylase, and carboxypeptidase, that any large-scale experimentation was carried out in this field. One of the problems was the scarcity of commercially prepared polymers which could be used successfully to immobilize enzymes; most were synthetic hydrophobic resins. To put these discoveries into the perspective of biochemical knowledge of the time, it must be remembered that the structure of proteins was a field in its infancy, the nature of DNA had just been discovered, biochemical genetics had not yet been invented, and few convincing hypotheses for the structure of biological membranes had been advanced. The discovery that the nature of the polymeric support used influenced the apparent kinetics of the immobilized enzyme must have been not a little disquieting to those who had hoped to solve the problems of cellular organization and control by studying such enzymes. The other major problem was that a lot of the work of this early period was both too early and in the wrong place. For example, McLaren in the mid-1950s was studying the adsorption of enzymes on to kaolinite. His considerable contribution at that time went largely unnoticed, because he was several years ahead of his time and his reports appeared in such journals as the *Annual Proceedings of the Society of Soil Science*, a journal not often consulted by biochemists!

2. The Importance of Hydrophilicity

It was not until the work of Manecke in 1961 and his observation that, in general, the level of activity of an immobilized enzyme depends on the degree of hydration of the polymer matrix, that polymers other than cellulose were investigated as supports to which enzymes could be covalently bonded. In 1963 Rimmon used a copolymer of L-leucine-*p*-amino-D,L-phenylalanine to immobilize chymotrypsin, papain, pepsin, and streptokinase. Manecke in 1962 pioneered the use of methacrylic acid-3-fluoro-4,6-dinitroanilide: methacrylic acid copolymer to immobilize pepsin, amylase, invertase, and alcohol dehydrogenase. This latter enzyme was the first oxido–reductase enzyme to be successfully immobilized, until that time all other immobilized enzymes were hydrolases.

Concurrently with this work the use of cellulose derivatives as supports for enzymes was investigated further. The continued popularity of cellulose was due to its inherent advantages; its high hydrophilicity, availability, potential for varied derivatization, and relative ease with which cellulose-based polymers can be produced either as particulate powders or as membranous films. Thus Mitz and Summaria in 1961 reported the coupling of trypsin and chymotrypsin to diazotized *p*-aminobenzoyl cellulose and also to the hydrazide derivative of carboxymethylcellulose. These two methods are still in use today.

3. Bridging the Gap

It is perhaps more useful when using cellulose as a polymer support not to have to build the reactive group into the cellulose (as in *p*-aminobenzoyl cellulose), but to use a chemical 'bridge' between the cellulose and the enzyme molecule. The requirement for this bridge molecule is that it must be small and, once reacted with the cellulose, have a group capable of reacting with the enzyme. Such a compound is cyanuric chloride (trichlorotriazine). This has three reactive C—Cl bonds (see Figure 3). One is first reacted with the cellulose (very rapidly), the second with the enzyme, and the third may be reacted with any convenient compound. Using this method Kay and co-workers (Kay *et al.*, 1968; Kay and Crook, 1967) linked galactosidase, lactate dehydrogenase, pyruvate kinase, and creatine kinase to cellulose in the form of filter paper. The particular advantage of cyanuric chloride as a bridge molecule is that the ionic nature of the enzyme–cellulose complex depends upon the ionic charge of the bridge molecule. This can be neutral, cationic, or anionic depending upon the nature of the compound added to the third C—Cl bond. Thus enzyme–cellulose complexes can be made polycationic by this method, a great advantage since most other methods tend to result in polyanionic complexes.

Another useful bridge molecule is glutaraldehyde. This contains two aldehyde groups at either end of a $(CH_2)_3$ unit. The aldehyde groups will react at neutral pH values with free amino groups. Thus one end of the glutaraldehyde molecule may be attached to the support, the other to the enzyme.

The most common activation method in use today is that involving cyanogen bromide (CNBr) (Axen *et al.*, 1967; Porath, 1974). The exact way in which this molecule reacts with cellulose has yet to be ascertained, but at high pH values it apparently reacts readily with the hydroxy groups of polysaccharides and the derivative will then react with free amino groups on the enzyme in mildly alkaline solutions. Problems are nevertheless experienced with this method (apart from the handling of cyanongen bromide) and it is becoming apparent that the bonding, particularly of small molecules, is not altogether stable.

4. Disadvantages of and Alternatives to Cellulose

Polysaccharides are not the ideal support material for enzyme immobilization as they suffer from two serious drawbacks. First, polysaccharides are susceptible to microbial attack. Nothing is more infuriating than finding a microbial growth devouring a precious immobilized enzyme preparation, particularly in a large-scale industrial application! Second, cellulose shows a high degree of non-specific adsorption of protein. Consequently, at the end of the preparative procedure the immobilized enzyme must be washed extensively in buffers of high ionic strength. This process may not only be expensive on a large scale but may often inactivate the enzyme, especially if it is dimeric or polymeric in its active form (see p.17).

Much work has thus been devoted to the search for polymer support materials that are hydrophilic but microbially non-degradable. Levin and Goldstein independantly reported, in 1964, the use of ethylene–maleic anhydride as a support for various enzymes. Other materials which have been successfully employed include glass (Weetal, 1969), and nylon (Inman and Hornby, 1972). A more general approach has arisen from the work of Inman and Dintzis who in 1969 pioneered the use of various derivatives of polyacrylamide. Many varieties of ready prepared acryl copolymers are now commercially available, the reactive groups of which commonly include diazo, aldehyde, carboxylmethyl, and hydrocyanate derivatives. A useful member of this group of polymers is a soluble acryl polymer derivative suitable for preparing soluble immobile enzymes.

III. COPOLYMERIZATION WITH MULTIFUNCTIONAL REAGENTS

Multifunctional reagents can be used not only to link enzyme molecules to cellulose or other polymers, but also to link enzyme molecules to each other. Although such a matrix may contain just enzyme molecules, it is usually in the interests of economy to copolymerize the enzyme with an inert protein such as albumin in order to increase the bulk of the final product.

The most commonly used multifunctional reagent is glutaraldehyde. Aldehydes in general and glutaraldehyde in particular have long been used by histologists as fixing agents; Baker in 1910 noted the protein gelling action of aldehydes. In fact, despite its age, Baker's observation is pertinent today, for it is extremely difficult to precipitate an enzyme matrix out of solution using glutaraldehyde, at best the solution usually gels. In order to produce an insoluble matrix of enzyme and glutaraldehyde it is necessary either to cause the glutaraldehyde to polymerize or to precipitate the enzyme (or adsorb it on to some insoluble surface). The net effect is either to increase the length of the bridging molecule or to decrease the intermolecular distance of the enzyme molecules. Using such approaches Richards (1964), Ogata (1968), Jansen (1969), and Habeeb (1967) have cross-linked and insolubilized carboxypeptidase A crystals, subtilisin novo, papain, and trypsin respectively.

Membranous sheets of cross-linked enzyme of controlled pore size can be prepared with glutaraldehyde. The enzyme is adsorbed on to a preformed cellulose nitrate membrane and then glutaraldehyde introduced to cross-link the enzyme molecules in place around the cellulose nitrate fibres. The cellulose nitrate can then be dissolved away with methanol leaving an enzyme membrane (Goldman *et al.*, 1968b).

Although many other multifunctional reagents have been used to cross-link enzymes, for example *N,N*′-bisdiazobenzidine-2,2′-disulphonic acid or 2,4-dinitro-3,5-difluorobenzene, only glutaraldehyde has found extensive use. This is probably because it reacts with proteins readily under mild conditions — 2,4-dinitro-3,5-difluorobenzene requires highly alkaline solutions to react and the presence of organic solvent to dissolve it — and is probably the least toxic of the bifunctional reagents used; *N,N'*-bisdiazobenzidine-2,2'-disulphonic acid is a potentially explosive carcinogen!

The major problem with this type of method is that the bifunctional reagent will often preferentially attack the active site of the enzyme, thus rendering it inactive. However, when it can be made to work, for example by reversibly blocking the active site, it works well.

IV PHYSICAL ADSORPTION

The oldest method of enzyme immobilization is that of physically adsorbing the enzyme on to a polymer matrix without covalent bonding. It is exceedingly easy to perform, the adsorbent and enzyme are stirred together for some time, but yields (enzyme bound per unit of adsorbent) are low and the enzyme is often partially or totally inactivated. A variety of adsorbents have been used; the binding forces may be ionic, hydrophobic, hydrogen bonds, or Van der Waals' interactions.

Unfortunately there is an inherent snag in the use of physical adsorption for enzyme immobilization. It is embodied in the principle 'what goes up must come down', that is the reversible nature of the bonding of enzyme to support may lead to desorption of the enzyme at a critical time. One factor which often causes desorption is the addition of substrate to the enzyme preparation. This is a particular hazard, for although other factors which might cause desorption (such as fluctuations in pH, temperature, or ionic strength) can be controlled,

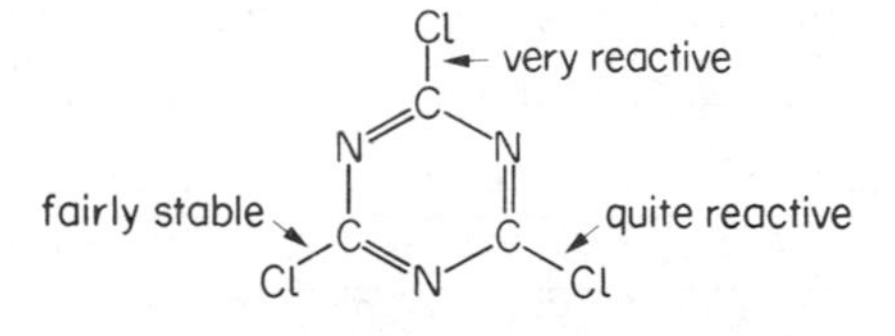

Cyanuric chloride

$OHC{\cdot}CH_2{\cdot}CH_2{\cdot}CH_2{\cdot}CHO$

Glutaraldehyde

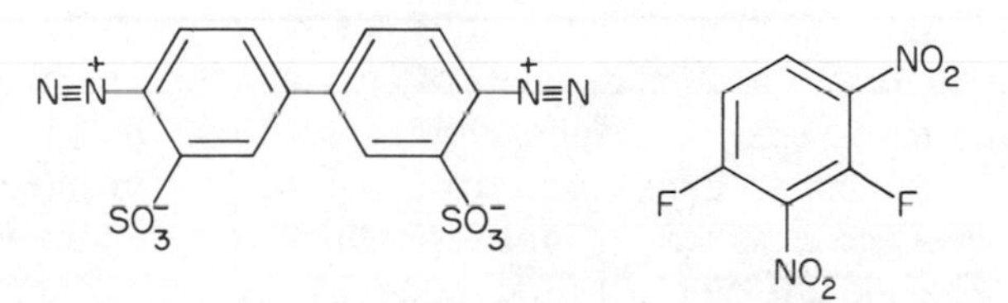

Figure 3. Multifunctional reagents

no enzyme can work without its substrate. It is interesting that despite this limitation it was just such a method that was used in the first commercial application of an immobilized enzyme.

The high adsorption of proteins by cellulose, which is a problem in the use of cellulose for the covalent bonding of enzymes, is here a great advantage. Cellulose based ion-exchange resins (e.g. carboxymethylcellulose and DEAE cellulose) have been extensively used and high adsorbent capacities have been demonstrated (up to 15% w/w protein : cellulose). Other materials often used, but with markedly inferior capacities, include polystyrene resins, kaolinite, collagen, alumina, silica gel, and glass.

One interesting development of the technique of immobilization by adsorption is in the use of an effector or activator of an enzyme, itself attached to a water-insoluble polymer, to bind that enzyme. For example, Fukui *et al.* (1975) demonstrated the immobilization of tyrosinase and tryptophanase by adsorption on to an insoluble derivative of pyridoxal-5′phosphate, an activator of these enzymes. This method has the added advantage that the enzyme is not only specifically adsorbed on to the polymer, but is also activated by the same process.

V ENTRAPMENT AND OCCLUSION

The entrapment of an enzyme molecule can be achieved in one of three ways.

1. Inclusion within the matrix of a highly cross-linked polymer.
2. Separation from the bulk phase by a semipermeable ‘microcapsule’.
3. Dissolution in a distinct non-aqueous phase.

An important feature of the first two types of method in this group (1 and 2 above) is that the enzyme is not actually attached to anything. There are therefore none of the steric problems associated with covalently or electrostatically binding an enzyme on to a polymer, for example binding the enzyme in such a way that its active site is obstructed by a portion of the polymer matrix (see p.21). In general, entrapment methods are performed by dissolving the enzyme in a solution of the chemicals required for synthesis of the enzyme phase and then treating this solution so that a distinct phase is formed. Cross-linked polyacrylamide gels may be formed by dissolving enzyme, acrylamide and methylene-bis-acrylamide in buffer and initiating polymerisation. The gel formed is mechanically disrupted to form small, enzyme loaded particles (Figure 4). Semipermeable microcapsules of nylon, of defined size and porosity, can be formed around droplets of enzyme solution. Alternatively, an aqueous solution of enzyme dispersed in a solution of PVC or cellulose triacetate may be extruded to form fibres containing droplets of enzyme. Liposomes, concentric spheres of phospholipid bilayers, may be used to encapsulate enzyme dissolved in the aqueous compartments between the bilayers. The medical importance of such preparations will be discussed later (p.64). Phospholipid bilayers have also been used as a lipid membrane in which the enzyme isomaltase/sucrase is ‘dissolved’ (Storelli, *et al.* 1972). This latter type of enzyme immobilization is obviously only suitable to enzymes that are normally incorporated into biological membranes and the use of the technique is really confined to a study of reconstituted membrane transport systems (see p.117).

VI CHOICE OF IMMOBILIZATION METHOD

The art of enzyme immobilization lies in knowing which method to choose. Choice is governed by a number of factors some of which will not be apparent until the procedure is tried. The initial choice is usually purely empirical; must the enzyme be attached to a specific type of support material or will any material do so long as the resultant immobilized enzyme preparation has high activity? Obviously if the former applies the choice is immediately limited.

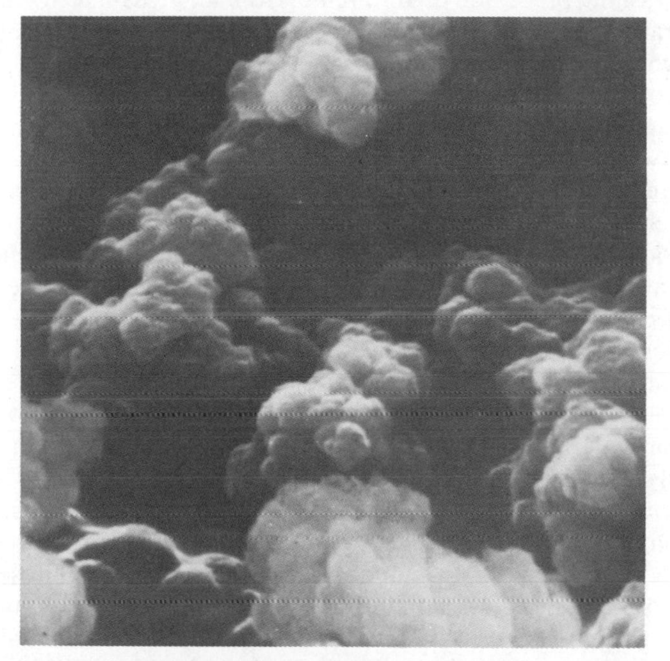

Figure 4. Scanning electron micrograph of highly cross-linked polyacrylamide gel particles. ($\times$ 8000)

Assuming that any support material will do, then the most immediate limitation is whether the immobilization procedure will inactivate the enzyme and whether the immobilized enzyme will actually work under the conditions to be employed. For successful immobilization, therefore, the following factors should, if possible, be considered.

1. The enzyme must be stable under the conditions required for reaction. For example, reactions requiring molar sodium hydroxide at 60 °C are best avoided.

2. If possible, cross-linking reagents should react preferentially with chemical groups other than those at the active site.

3. If the above condition cannot be met (as is usually the case) the cross-linking reagent should be as large as possible to prevent it from penetrating the active site. A polymer such as an 'activated' cellulose is thus preferable to a small bifunctional reagent (e.g. glutaraldehyde).

4. Where it is feasible, the active site of the enzyme should be protected in some way. This may be achieved in sulphydryl enzymes by first reacting the enzyme with glutathione or cystine then reactivating the enzyme once it is immobilized. Certain enzymes, for example papain, may be immobilized in the inactivated state, for other enzymes blocking the active site can sometimes be successfully achieved by incorporating saturating concentrations of substrate in the reaction mixture.

5. The washing procedure used to remove the un-cross-linked enzyme from the preparation must not adversely affect the enzyme. Cellulose polymers will usually strongly adsorb the enzyme such that high ionic strength solutions are required to wash the immobilized enzyme. Thus the use of cellulose as a support material for a polymeric enzyme is sometimes inadvisable, as the process of washing may cause the enzyme to dissociate.

6. If the enzyme, once immobilized, is to be used as a high efficiency catalyst in some chemical reaction (rather than to be studied in its own right) consideration must be given to the nature of the reaction before choosing the immobilization method. For example, it would be pointless selecting the process of physical entrapment within a gel matrix if the enzyme is required to catalyse the breakdown of a high molecular weight polymer such as a polysaccharide. Equally a polyanion would be of limited use as a support for an enzyme catalysing the conversion of an anionic substrate into a cationic product particularly if the enzyme suffers from product inhibition (see p.46).

7. Finally, the mechanical properties, in particular mechanical stability and physical form, of the support material must be considered. The immobilized enzyme may be required as a membranous sheet, thus it must obviously be possible to form the support material into a membrane. Cellulose polymers (e.g. cellulose acetate or cellulose nitrate) are perhaps the easiest materials to form into sheets; however with a little ingenuity even materials such as collagen or highly cross-linked polyacrylamide (normally a floccular substance) can be formed into membranes. The mechanical properties of certain particulate immobilized enzyme preparations make them unsuitable for use in large-scale reactor systems, others, however, may be specifically designed for use in a reactor (see the use of magnetic iron p.83).

The discovery of an immobilization method that satisfies all of these requirements is as might be expected a rare event; compromise is the usual order of the day.

Chapter 2

Effect of Immobilization on Enzyme Activity

Much has been written in recent years on the effect of immobilization on an enzyme's activity, from both an experimental and theoretical point of view. The theoretical mathematical analysis of even simple, well defined, immobilized enzyme systems is extremely complex and so we shall attempt here to deal with the matter in a purely descriptive fashion. Once the concepts involved have been thus grasped, full mathematical analysis becomes less formidable.

Enzymes, like people, are affected by the surroundings in which they find themselves. Surrounding an enzyme with a high concentration of insoluble polymer is similar to locking a man in a room. Just what the enzyme or the man can do and how efficiently it or he can do it, will depend entirely on what is allowed into or out of the environment and what physical constraints are present. Chain the man to the wall, put him in a room 1.5 × 1.5 × 1.5 m, allow no food to pass to him, and he will be capable of very little. So it is with the immobilized enzyme. The polymer matrix to which the enzyme is attached may prevent substrate from passing to the enzyme (or cause product accumulation) or may effectively 'chain' the enzyme thus preventing sufficient movement for it to perform its task. This chapter will consider these effects separately, first dealing with the way in which the polymer matrix may influence the microenvironment surrounding an enzyme (I), then considering the way in which the polymer might exert constraints directly on the enzyme molecule (II), and finally explaining some of the effects such influences may have on the enzyme's apparent behaviour (III) and (IV).

It would perhaps be useful to define here certain terms that will be used. Enzyme molecules are affected by events (e.g. pH changes) occurring in their immediate vicinity. For an enzyme in dilute homogeneous solution the events occurring in the vicinity of an enzyme molecule are the same throughout the solution, in particular at those points in the solution where these events are measured. For an immobilized enzyme, however, this is not necessarily the case. The polymer matrix provides the enzyme with a microenvironment which may differ in many respects (e.g. pH, substrate concentration) from the solvent in which the immobilized enzyme is suspended. To distinguish it from the microenvironment of the enzyme, the solvent is referred to as the 'bulk phase', 'bulk solution', or 'macroenvironment'. It is, of course, from the bulk phase that measurements of pH, substrate concentration etc. are taken (see Figure 1).

I THE MICROENVIRONMENT

There are two distinct ways in which a polymer support might affect the microenvironment surrounding an immobilized enzyme. The first may be considered to be a partitioning effect. By virtue of its own physical chemistry, the polymer may attract (or repel) substrate, product, inhibitor, or other molecules to its surface thus concentrating (or depleting) them in the immediate vicinity of the enzyme. The second way in which the polymer may affect the enzyme's microenvironment may be by presenting itself as a barrier to the free diffusion of molecules both to and from the enzyme. Either partitioning or diffusion limitation effects may be present on their own in a given immobilized enzyme system or they may both be present, acting either synergistically or antagonistically (see Figure 5).

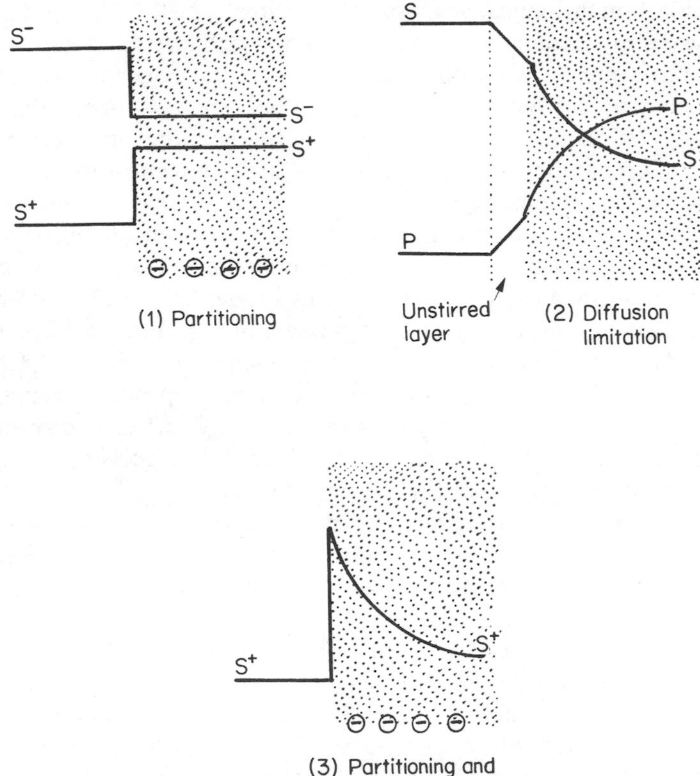

Figure 5. The effect of partitioning and diffusion limitation on solute concentration profiles in immobilized enzyme particles

1. Partitioning

Let us first consider the effect of partitioning. The most obvious example here would be the case of an enzyme, acting on an cationic substrate, immobilized by attachment to a polyanion (i.e. a negatively charged polymer). The positive substrate would be concentrated around the negative polymer and thus, although the average concentration of the substrate in the system might be low, around the enzyme it will be relatively high. Of course the polymer would not only attract the positively charged substrate, but any other cations, for example protons. Thus one might expect that, in this case, the hydrogen ion concentration around the enzyme will be higher (and the pH therefore lower) than the 'average' concentration throughout the system. The converse will obviously apply when the polymer and solute bear the same charge. It is quite easy therefore to visualize the situation where, given a polyionic polymer support, no ionic solute will be homogeneously distributed throughout the system, but will be present in the microenvironment around the enzyme at a concentration different to that in the bulk phase. This is really the crux of the matter, for in assigning values to substrate, proton, product, or other solute concentrations in enzyme rate equations, the values used are the values measured from the bulk phase, which may or not be the same as the values in the enzyme's microenvironment. Clearly then, the application of dilute solution Michaelis–Menten type kinetics to an immobilized enzyme system is bound to produce strange results if partitioning effects are present.

An odd variant of this type of partitioning effect has been described by Goldman *et al.* (1968b), who immoblized papain in a cellulose nitrate membrane. When acting on gelatin the immobilized papain, although less active than papain in dilute solution, was able to digest the gelatin more thoroughly (Figure 6). Their explanation was that the cellulose nitrate of the membrane was adsorbing the gelatin and so causing it to denature, thus allowing it to be more easily digested (see p.20 for alternative explanation).

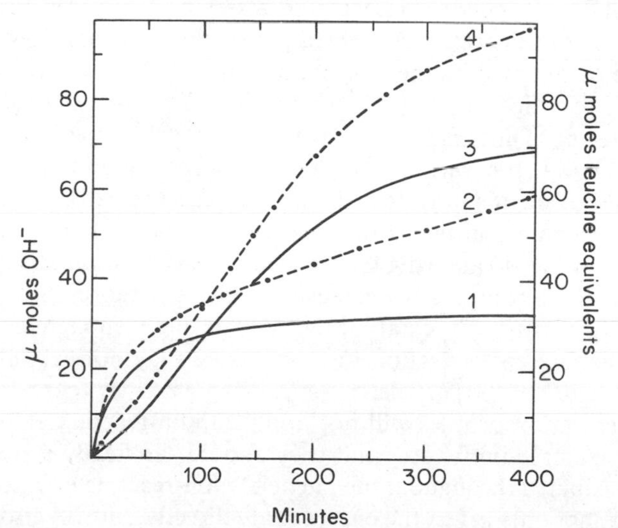

Figure 6. Time course of gelatin digestion by papain in dilute solution (curves 1 and 2) and as a papain membrane (3 and 4) expressed as alkali required to titrate the reaction products (solid lines) and free amino group content of the reaction mixture — leucine equivalents (broken lines) Reprinted with permission from Goldman *et al.*, *Biochemistry*, 7, 486 (1968). Copyright 1968 American Chemical Society

Although ionic partitioning is the most commonly observed form other types are quite possible, for example hydrophobic interactions between solute and polymer support. The possibility of specific interactions between the support and a particular solute also exists.

2. Diffusion Limitation

Whereas partition effects depend ultimately upon the physical chemistry of the polymer and the solute in question, diffusional limitation is largely a matter of physical size. It should be quite apparent that if the pore diameter of the polymer matrix is smaller than the substrate molecule, so preventing access to the enzyme, then no reaction will take place. However, diffusion effects are seldom so simply determined.

Unlike partition effects, except in the extreme example just given, diffusional effects will only result in concentration differences between the microenvironment and the bulk solution if the enzyme is active. It is the alteration of substrate and product concentrations in the enzyme's microenvironment, as a result of the enzyme's catalytic activity, that sets up the concentration gradients of substrate into and products out of the immobilized enzyme particle. Such concentration gradients are a result of the slow rate of diffusion of molecules in gels and unstirred solutions and the high catalytic activity of most enzymes.

When immobilized enzyme and substrate are first mixed, the substrate diffuses into the particle and the enzyme starts to utilize it at a rate dependent upon the substrate's concentration. Thus a concentration gradient is rapidly set up throughout the polymer matrix and a steady state is reached where the rate of diffusion of substrate at a defined point within the polymer particle will equal its rate of removal by the enzyme at that point. The further away this point is from the edge of the polymer particle the lower the substrate concentration must be. Thus, as a direct result of this, the deeper into the polymer particle, the lower the rate of substrate utilization and hence substrate diffusion. As the rate of substrate diffusion decreases towards the centre of the particle the substrate concentration gradient cannot be linear, but must be curved (see Figure 7), so allowing the substrate concentration difference over a given distance to be continually decreasing with penetration into the particle. Obviously if the enzyme's catalytic activity is high enough and the bulk solution's substrate concentration low, substrate may not actually penetrate right into the centre of the particle.

Such concentration profiles will not apply to non-reacting effectors of the enzyme, for example inhibitors, and they should, at steady state, be homogeneously distributed throughout the particle. Non-reacting effector molecules will, however, not only affect the enzyme rate directly but will also, as a result of affecting the enzyme rate, affect the concentration profiles of substrate and product. Introduction of an inhibitor into a reacting immobilized enzyme system, in which the bulk solution substrate concentration is low such that

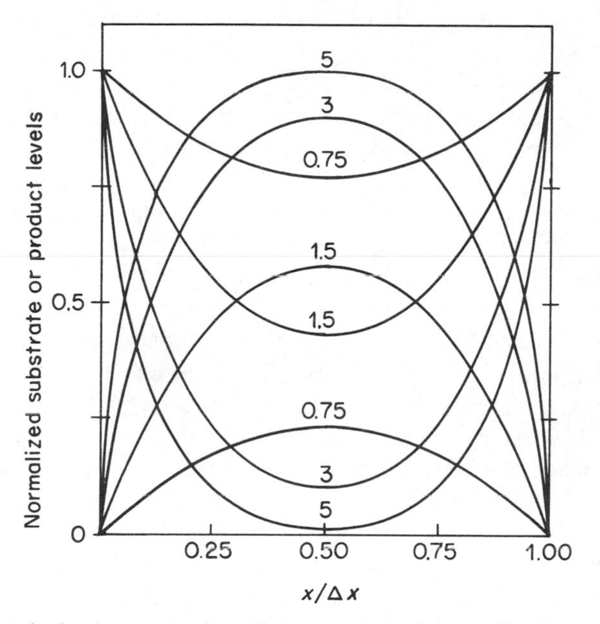

Figure 7. Theoretical substrate and product concentration profiles in reacting enzyme–membrane system at different values of diffusional resistance of the membrane for a fixed intrinsic enzyme activity. $X/\Delta X$ represents the thickness of the membrane. Substrate concentration profiles are concave upwards, product convex upwards. Reprinted with permission from Goldman *et al.*, *Biochemistry*, 7, 4518 (1968). Copyright 1968 American Chemical Society

substrate does not penetrate all of the particle, will cause the enzyme's catalytic activity to be reduced, thus substrate diffuses deeper into the particle and hence has access to more enzyme molecules. This will obviously moderate the effect of the inhibitor. An effect of this kind might explain why certain immobilized enzyme preparations are less sensitive to changes in pH than the native enzyme would appear to be (see p.30). The converse may apply to an enzyme activator.

Diffusional effects may be divided into two types, the external diffusion barrier and the internal diffusion barrier. The external diffusion barrier is a result of the thin, unstirred layer of solvent that surrounds the polymer particle, the so-called Nernst layer. Solutes diffuse in this layer by a combination of passive molecular diffusion and convection. The thickness of this layer is affected, within limits, by the speed at which the solvent around the immobilized enzyme particle is stirred. Increasing the stirring rate will reduce this external diffusion barrier. Thus an important factor in any experiment involving an immobilized enzyme preparation is the stirring speed.

Internal diffusion limitations, as their name implies, are limitations to free diffusion within the polymer matrix imposed by the polymer matrix. Within the polymer matrix diffusion takes place by passive molecular diffusion only, and is not affected by stirring speed. Internal diffusion effects will be more marked if the enzyme is immobilized by entrapment within the polymer matrix rather than attachment to the surface of the polymer matrix. Whether the overall rate of diffusion of solute into (or out of) the immobilized enzyme particle is limited by the internal or external diffusion barrier will depend upon the particular system involved. In most instances either the external or internal limitation may be ignored as only one can be rate-limiting.

In attempting to discover why a particular immobilized enzyme system behaves as it does, it is important to be able to distinguish between internal and external diffusion effects and partitioning effects. In general, if only partitioning effects are present then, at a constant ionic strength, the immobilized enzyme will obey Michaelis–Menten-type kinetics, although the measured kinetic parameters are unlikely to be the same for the immobilized as for the dilute solution enzyme. If diffusional effects are present then plotting typical Lineweaver–Burk reciprocal plots is likely to produce curves which are often sigmoidal in shape. The significance of this is discussed later. Various devices, either experimental or theoretical, may be employed in order to eliminate any effects due to diffusion barriers or partitioning. This will leave the intrinsic kinetic parameters of the enzyme which may not be the same as those of the enzyme in free solution. The cumulative effect of diffusional limitations, partioning, and direct effects of immobilization on an enzyme's kinetic parameters are summarized below.

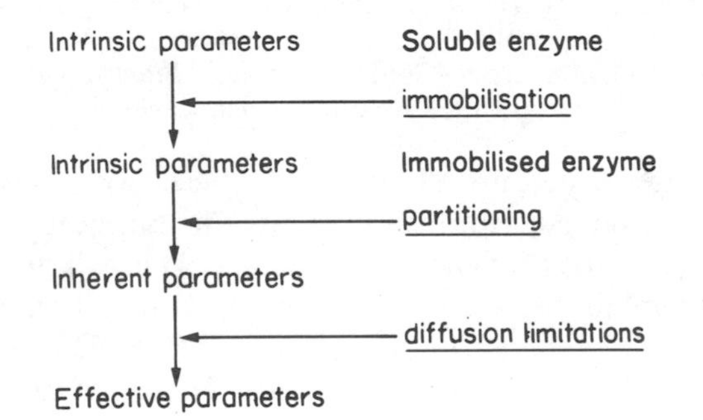

II THE EFFECT ON THE ENZYME MOLECULE

We have seen in the previous section that the process of immobilization affects the apparent kinetic parameters of the enzyme in a number of stages. The first change that may take place is a direct effect of immobilization on the enzyme molecule itself altering the intrinsic properties of the enzyme. There are a number of ways in which the immobilization process may directly affect the enzyme and these are dealt with in turn. It should be noted that any, all, or none of these effects may be present in a particular system depending upon the nature of the immobilization process, the enzyme, and the reactants.

1. Conformational Changes

It is quite feasible that if the enzyme is immobilized by covalent attachment to the polymer matrix, constraints may be applied to the conformation of the enzyme and its ability to change.

It is generally realized that the ability of an enzyme to act as a specific catalyst will depend primarily upon two factors. First, the enzyme must be in precisely the correct conformation. Second, it must be able to change its conformation during catalysis. The first requirement is exhibited by the substrate specificity of most enzymes and by the fact that partial denaturation may lead to inactivation. The second requirement is seen in the 'Induced Fit' mechanism of enzyme–substrate binding and in the mode of action of allosteric enzymes.

Before we consider each of these requirements a word about the polymer matrix. Many polymers, particularly those that have a gel-like structure, may themselves undergo conformational changes if their environment is altered. For example polyacrylamide gels take up to 2 weeks to hydrate fully and during this time their conformation is continually changing. We shall assume here, however, that the polymer structure is invariant. For a discussion of the effect of changing polymer conformation see Mechanocatalysis (Chapter 4, Section I).

The most obvious effect of immoblizing an enzyme on to a rigid support will be that the enzyme's structure will be held in one position if there exist multiple attachment points between the enzyme and polymer. If the enzyme is dependent for its catalytic activity upon conformational changes then one might expect to observe perturbations in the values of K_m and/or V_{max}. The situation will be more complex in the case of a polymeric enzyme and here a host of problems may be encountered. Let us take the example of a dimeric, allosteric enzyme. The first problem that may be encountered is that the enzyme may be immobilized via one subunit only. This has a number of implications.

1. The immobilized subunit may be totally inactivated (see below), as a result of which the enzyme will no longer behave in an allosteric fashion.

2. The washing procedure may dissociate the enzyme dimer, resulting in an immobilized monomeric enzyme whose kinetic parameters will naturally be quite divorced from the native enzyme.

3. Restriction of conformational changes in the enzyme may cause it to loose its allosteric properties.

4. Immobilization may not restrict conformational changes within each monomer but may affect the transmission of conformational change from one subunit to the other.

5. While immobilization may not affect the ability of the enzyme to change its conformation or bind its substrates, it may alter or render inaccessible any activator or inhibitor binding sites. Much work has been performed in recent years directly measuring by physical methods the conformational state of immobilized enzymes and this has confirmed the hypothesis that the conformation of the enzyme may indeed be altered by immobilization.

2. Stability

Early reports on the stability of immobilized enzymes often demonstrated an apparent enhancement of the enzyme's stability. It was suggested that this could be due to two reasons. First, the immobilization of proteases would effectively prevent autolysis by restricting intermolecular contact. Second, fixing an enzyme to a polymer matrix which is unaffected by heat or pH changes was thought to provide the enzyme with a protective shell preventing it from altering its conformation in response to such changes. While the first observation and its interpretation has been repeatedly confirmed, the second has come under increasing attack in recent years.

It is possible that in a limited number of cases immobilization may enhance the conformational stability of an enzyme. It has been observed that trypsin immobilized on EMA remains active in 6 mol l^{-1} urea, when the soluble enzyme is completely denatured. However this kind of stabilization does seem to be the exception rather than the rule.

How then was the hypothesis of enzyme stabilization generated? To answer this question we must examine the manner in which immobilized enzyme stability is studied. The usual technique is to subject both the immobilized and free enzymes to long periods of heat treatment and to assay repeatedly the activity under standard conditions. Such experiments frequently produced results of the pattern shown in Figure 8. From a superficial examination of these results it would appear that the immobilized enzyme is indeed more stable to heat treatment. However, two further factors must be taken into account, protein concentration and substrate diffusion limitation.

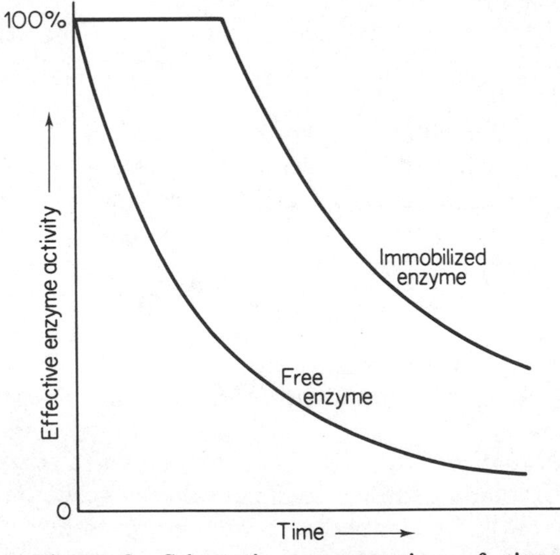

Figure 8. Schematic representation of time-dependent stability of immobilized (1) and free solution enzyme (2)

a. Safety in Numbers

In recent years it has become apparent that the stability of an enzyme in free solution is not just dependent upon the physical conditions in which it is placed, but also depends upon the enzyme or even total protein concentration. Thus, while an enzyme may be relatively unstable in the dilute solution in which it is usually studied, raising the enzyme's concentration can enhance its stability. This of course sheds considerable doubt on the validity of comparisons of free and immobilized enzyme stabilities, where the free enzyme is usually in very dilute solution while the local concentration of the immobilized enzyme may be relatively high. Few studies have actually been performed taking such considerations into account and, of those studies that have been performed, it would seem that stabilization due to *per se* immobilization is the exception.

b. The 'Zulu' Effect

The principal strength of the Zulu armies of the nineteenth century was not so much in the military skill of individuals to avoid death, but in the fact that for every 100 killed there were another 1,000 to take their place. Immobilized enzymes can behave in very similar manner. The high loading of most immobilized enzyme preparations means that under normal conditions only a fraction of the enzyme may be involved in the reaction. Consider an immobilized enzyme particle with a high local enzyme concentration. It is probable that with normal substrate concentrations, the substrate will only be able to penetrate a short distance into the particle before it is all converted to product. This means that the enzyme at the centre of the particle will not take part in the reaction. Even if 50% of the enzyme is inactivated (e.g. by heat treatment) there may still be sufficient enzyme present to remove all the substrate penetrating the particle. Thus the catalytic activity of the particle will appear unchanged because the reaction is in effect limited by the rate of substrate diffusion rather than the intrinsic enzyme activity (Figure 9). This also explains the observation that the immobilized enzyme may apparently remain active for a period of time and then its activity will suddenly decrease (when the active enzyme concentration drops below the point at which substrate diffusion is the limiting factor).

3. Steric Restrictions

Immobilization will prevent direct contact between enzyme molecules and, by enclosing the enzyme in a polymeric cage, it may prevent access of substrate/effectors to the enzyme.

The prevention of contact between enzyme molecules will only have practical significance in situations where inter-enzyme contact produces a kinetically observable event. The most important example of this is the

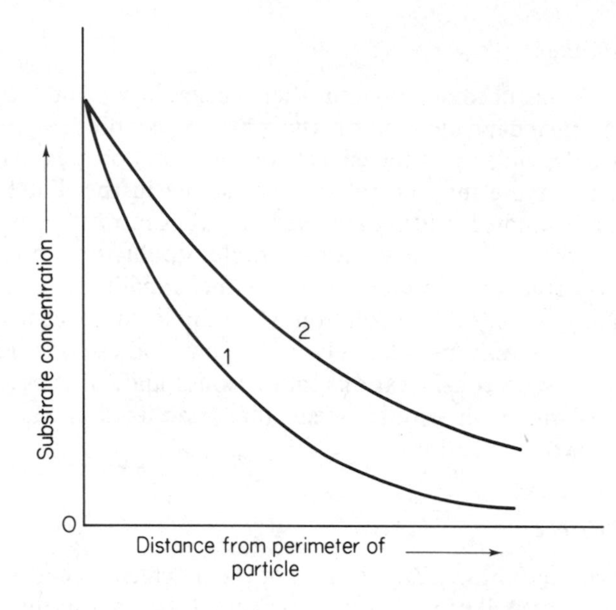

Figure 9. Schematic representation of the effect of intrinsic enzyme activity on the penetration of substrate into immobilized enzyme particles. (1) High enzyme activity (2) Low enzyme activity

autolysis observed in solutions of proteolytic enzymes, e.g. papain. In the example mentioned on page 13 of a papain/cellulose nitrate membrane used to digest gelatin, the observation that the papain membrane digested more of the gelatin than the dilute solution enzyme (Figure 6) might be explained by the proposal that the papain membrane does not suffer from autodigestion of the papain and will thus remain active for a longer time period. It can be noted, by way of a diversion, that multiple explanations of the observed behaviour of immobilized enzymes are usually possible but unfortunately not always considered.

If an enzyme is buried inside a polymer matrix then the smaller the substrate/effector molecule, the less chance there is that the molecule's diffusion to the enzyme will be hindered by the matrix. Immobilized proteolytic enzymes are particularly useful in the study of gross steric effects because they will operate on substrates with molecular weights ranging from a few hundred to several hundred thousand. When trypsin is attached to ethylene–maleic anhydride it will rapidly catalyse the hydrolysis of benzoyl arginine ethyl ester (BAEE) and catalyse the hydrolysis of casein more slowly. The low molecular weight (*c.* 6,500) pancreatic trypsin inhibitor will inhibit the action of trypsin on all its substrates. Soya bean trypsin inhibitor, which has a high molecular weight (21,000) will inhibit the action of trypsin on casein but not on BAEE or small peptides. The trypsin buried deep within the polymer matrix is accessible to small substrate/effector molecules while large molecules can only come into contact with the trypsin on the very outside of the particle. Finally we must consider the 'wrong way round effect'. When an enzyme is bound to a polymer there is no guarantee that it will bind the correct way round and, often as not, it is likely to be bound with its active site blocked by the polymer matrix.

4. Inactivation

Little needs to be said about inactivation of the enzyme during the immobilization procedure. It is sufficient to point out that those immobilization processes that rely upon harsh reaction conditions (e.g. high pH, presence of free radicals, oxidizing agents) are most likely to inactivate some or all of the enzyme to be immobilized. Thus the final specific activity of the immobilized enzyme may be substantially less than that of the free solution enzyme. This effect is exacerbated by the hypereactivity of groups at the enzyme's active site, so that the enzyme may be inactivated without any conformational change. A composite drawing of all of the possible effects of immobilization on an enzyme is given in Figure 10.

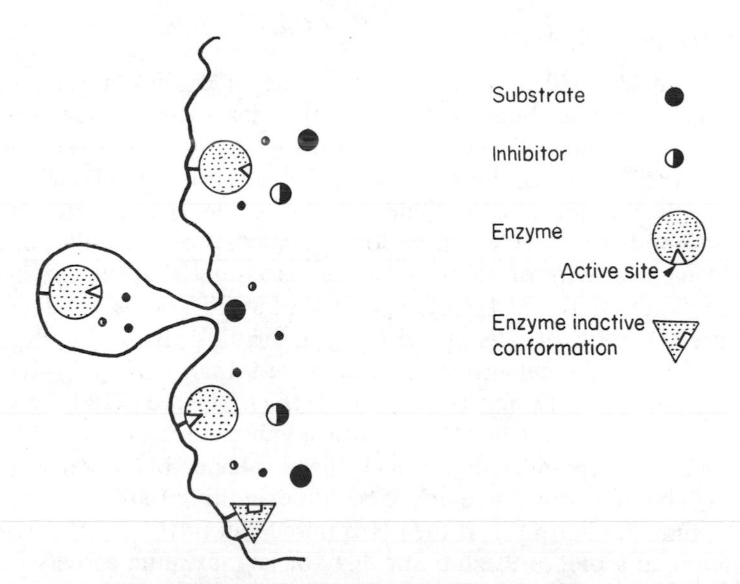

Figure 10. Illustration of possible steric effects of the polymer matrix on immobilized enzyme

III PERTURBATION OF ENZYME pH DEPENDENCE

The nature of an enzyme's dependence upon pH has been extensively observed and documented. In most of these studies the enzyme has been isolated, purified, and then dissolved and diluted in a buffered aqueous solvent. If, however, an enzyme is observed in a natural or artificial environment that mimics its state *in vivo*, that is in a heterogeneous environment associated with polymer matrices, then dramatic changes in the enzyme's apparent response to pH may take place. The study of such changes and the factors influencing them are of great interest not only to those wishing to learn more of the way in which enzymes work *in vivo*, but also to the biochemical engineer who also needs to be aware of all the factors that may influence the activity of an immobilized enzyme preparation. Most of the effects of perturbation of pH dependence observed in immobilized enzyme preparations may be explained by considering the distribution of protons throughout the system and analysing the factors that might lead to relative accumulation or depletion of protons in the microenvironment around the enzyme. Thus one is considering variations in the enzyme's inherent or effective behaviour (see p.16). Explanations based on consideration of changes in the enzyme's intrinsic behaviour were discussed in Section II of this chapter.

Two factors may give rise to heterogeneous distribution of protons in an immobilized enzyme system; partitioning effects by polyionic matrices and the restriction of proton diffusion by the polymer matrix. These we shall deal with separately.

1. Partitioning of Protons

Polyionic matrices will have the general effect of causing a partitioning of protons between the bulk phase and the enzyme's microenvironment. Polyanions will tend to concentrate protons (thus lowering pH) around the enzyme while polycations will tend to expel protons (raising pH) (see Figure 5). Let us take the example of an enzyme immobilized on to a polyanionic matrix. The pH around the enzyme will be lower than that of the bulk phase from which the measurement of pH is taken. Let us assume that the enzyme, when studied in dilute solution, has a pH optimum of 8 and a bell-shaped pH profile (see Figure 11). With an external pH (pH_e) in the bulk phase of 8, the internal pH (pH_i) of the microenvironment will be less than 8 (e.g. 7) due to the partitioning effect of the matrix. Thus while the measured pH of the system is 8 the enzyme is operating at pH 7, at which value it exhibits only 50% of its maximum activity (point *A*, Figure 11). If the value of pH_e is raised to 9, the enzyme will be functioning at a pH_i of 8, thus exhibiting 100% of its maximum activity (point *B*, Figure 11). If the pH is raised still further to 10, the enzyme will function at a pH_i of 9 exhibiting 50% of its maximum activity (point *C*, Figure 11). Hence the immobilized enzyme will appear to shift its pH/activity profile upwards by 1 pH unit.

Such displacement of the pH optimum in immobilized enzyme systems has been frequently observed. Goldstein (1972) observed a shift of 1 pH unit towards alkaline values of the pH optimum of chymotrypsin acting on acetyl-L-tyrosine ethyl ester, when immobilized on the polyanion ethylene–maleic anhydride copolymer. Conversely they observed a downward shift of the pH optimum by 1 unit when chymotrypsin was immobilized on the polycation polyornithine (see Figure 12).

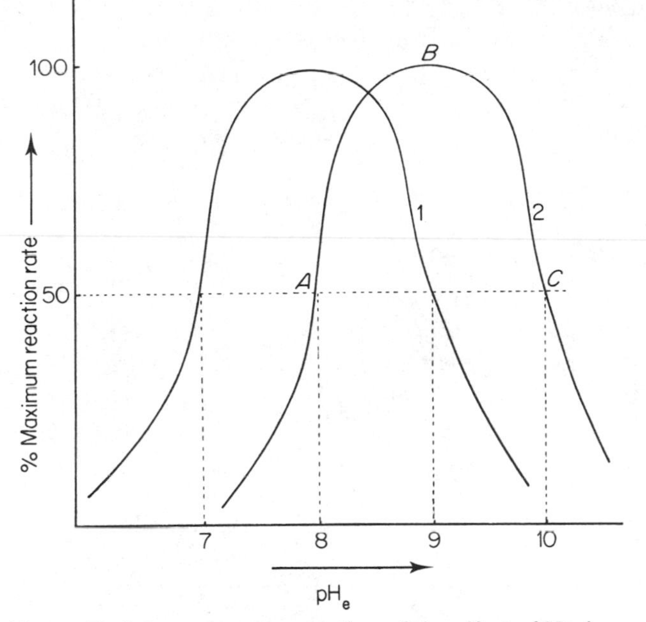

Figure 11. Schematic representation of the effect of H^+ ion partitioning by the polymer matrix of an immobilized enzyme (2) compared to the same enzyme in dilute solution (1). The matrix is a polyanion

Such displacements of the pH/activity profiles depend upon partitioning of protons effected by the presence of ionized groups on the polymer matrix. Thus the addition of solutions of high ionic strength diminish any displacement by covering up the ionized groups. Similarly, high concentrations of buffer also destroy these effects. Such perturbations are therefore only seen in solutions of low ionic strength containing little buffer. The moderating effect of buffer has been clearly demonstrated by Goldman *et al.* (1968b), who showed that when buffer was forced through a cellulose-nitrate membrane to which papain had been attached with *N,N'* bisdiazobenzidine-2,2'-disulphonic acid, the perturbation of the pH/activity profile virtually disappeared (Figure 13).

The equal displacement of both limbs of the pH/activity profile may be explained by considering the nature of the proton partitioning. An ionic partition coefficient (Δ) is defined as the ratio of the ionic concentrations in both phases, thus

$$\Delta_{H^+} = \frac{H_i^+}{H_e^+}$$

Where H_i^+ = concentration of protons in the matrix
H_e^+ = concentration of protons in the bulk phase

This takes no account of changes in the ionic activity as enzymes have been

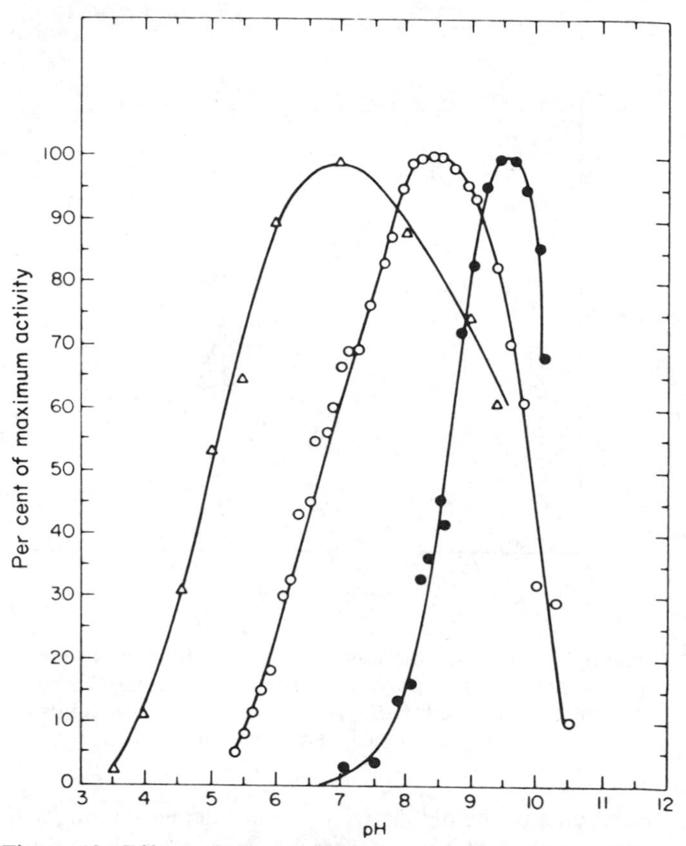

Figure 12. Effect of the polymer matrix on the pH profile of chymotrypsin, in dilute solution (○), immobilized to polyanionic ethylene–maleic anhydride copolymer (●), and immobilized to polycationic polyornithine (△). Substrate acetyl-L-tyrosine ethyl ester, ionic strength 0.01. Reprinted with permission from Goldstein, *Biochemistry*, **11**, 4072 (1972). Copyright 1972 American Chemical Society

shown to respond to proton concentration. Δ_{H^+} is also a function of the density of charge on the polymer matrix and the ionic strength of the solution. At very low proton concentrations (above pH 3) changes in pH will have a negligible effect on the ionic strength of the solution and thus will not affect the value of Δ_{H^+}. Hence the ratio of H_i^+/H_e^+ remains constant over a wide range of pH and it follows therefore that the value of $(pH_i - pH_e)$ will also remain constant.

2. Limitation of Proton Diffusion

The polymer matrix to which an enzyme has been immobilized can act as a barrier to the free diffusion of protons.

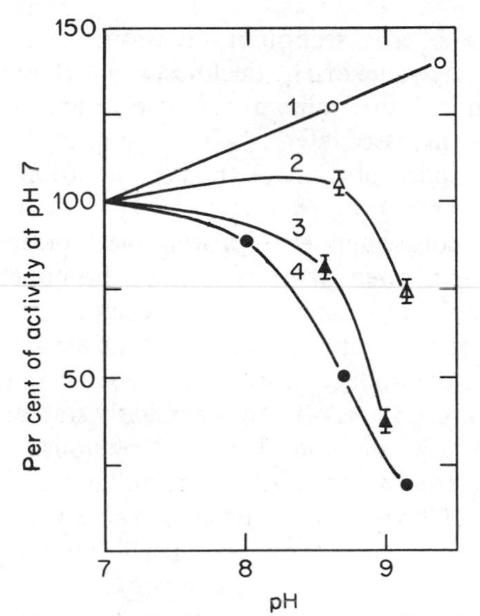

Figure 13. Effect of buffer on pH profile of papain membrane acting on benzoyl arginine ethyl ester in the absence of buffer (1), in the presence of buffer (2), and with buffered substrate forced through the membrane at 5 atm (3). Curve (4) is the pH profile of papain in free solution. Reprinted with permission from Goldman *et al.*, *Biochemistry*, **7**, 486 (1968). Copyright 1968 American Chemical Society

In dilute solution protons will diffuse up to 100 times faster than most other monovalent ions. This is possible partly because of the small size of the proton and partly because of its ability to hop from one water molecule dipole to the next. If in an immobilized enzyme system the proton concentration were of the same order of magnitude as the other solute concentrations, then the limitation in the free diffusion of protons would be negligible compared to the effect on substrate and product molecules. However substrate concentrations are usually greater than 10^{-4} mol l^{-1} whilst proton concentrations are frequently less than 10^{-7} mol l^{-1}. Thus restriction of proton diffusion may become apparent under conditions when no significant effect is observed on substrate diffusion. It is also worth noting at this stage that the restriction on proton diffusion will become more noticeable as the pH rises. This effect is most easily explained by examining Ficks Law of diffusion.

$$J_{H^+} = \frac{A.D.(H_i^+ - H_e^+)}{d}$$

Where J_{H^+} = rate of proton flux (diffusion),
D = diffusion coefficient,
d = path length of diffusion,

$A \quad$ = Area of cross section of diffusion.

Clearly, as the absolute value of H_i^+ declines, the difference (pH_i—pH_e) must increase to maintain the same value of (H_i^+—H_e^+) and thus J_{H^+}. The further significance of this is discussed later.

The assumption implicit in most of the considerations of this book is that the enzyme reaction is being studied once the system has reached some form of equilibrium. This of course applies to proton distributions. Assuming that no partition effects occur then protons will be homogeneously distributed throughout the immobilized enzyme system, even if the matrix is capable of severe proton diffusion limitation, unless protons are involved in the enzyme reaction as substrate or product. Curiously, in biochemical texts the involvement of protons in a reaction is often not explicitly stated, despite the fact that such involvement is very common. The most obvious example of a reaction liberating protons is ester hydrolysis. Other similar reactions, for example all reactions involving ATP or NAD, are listed in Table 1. Occasionally the pH of the system will dictate whether the reaction will liberate (or utilize) protons; peptide bond hydrolysis will be essentially proton-utilizing at low values of pH, proton-liberating at high values of pH, and neither over a wide band of neutral pH values; the hydrolysis of urea is proton-liberating above pH 9.3 and proton-utilizing below this pH. It should be apparent when considering the effects of proton liberation by a reaction, that protons, unlike most other products of reaction, affect the enzyme's activity when present at very low concentrations.

Table 1 Acid utilizing or producing reactions

Reaction	Enzyme	Product Acid/Alkali
Hydrolase	Urease	Acid > pH 9.3 alkali < pH 9.3
Esterase	Trypsin	Acid
Peptidase	Trypsin	Neutral between pH 4 and 8
Dehydrogenase	Alcohol dehydrogenase	Acid if NADH is product
Phosphatase	Alkaline phosphatase	Acid
Kinase	Hexokinase	Acid > pH 6.8
Lysase	Aspartate ammonia lyase	Alkali

Consider the case of an enzyme, immobilized on to a non-ionic support, which catalyses a proton-liberating reaction in the absence of buffer. Once initiated, the reaction liberates protons which begin to accumulate in the polymer matrix. This raises the proton concentration in the matrix and thus provides the concentration gradient necessary for net outward diffusion of the protons liberated. A steady state will eventually be reached where the proton concentration difference between the matrix microenvironment and the bulk phase is sufficient to drive the protons out at a rate equal to their generation. However, the accumulation of protons is the microenvironment will lower the pH in the immediate vicinity of the enzyme. This may influence the rate of catalysis and hence the rate of proton production. If the pH before the reaction is initiated is above the pH optimum of the enzyme, then the liberation of protons and the consequent fall in pH_i will tend to raise the catalytic rate, increasing the rate of proton production. In order to achieve the steady state the rate of proton flux out of the matrix must now also increase. This will require a larger proton concentration gradient and consequently the microenvironment proton concentration will rise, lowering pH_i still further, causing a further increase in reaction rate and proton production. Such a system will therefore tend to be autocatalytic. The steady state will eventually be reached when a significant rise in proton concentration has a minimal effect on the enzyme reaction rate. The exact point at which this occurs will depend upon a number of factors, the nature of the enzyme's pH/activity profile — the steeper the curve the more pronounced the effect — the diffusion coefficient of protons within the polymer matrix, enzyme loading, and the extent of substrate and product diffusion limitation. An enzyme reaction occurring under such conditions will show a gradual acceleration of rate until the steady state is reached. It is possible that the pH_i around the enzyme may be lowered to the value of the enzyme's pH optimum. A further fall in pH would then tend to lower the reaction rate and thus a fall in pH to below the pH optimum is unlikely except in certain circumstances. It is possible, in a situation where proton diffusion is severely limited, that the maximum rate of proton diffusion at the pH of the enzyme's pH optimum may be less than the inherent rate of reaction. In such a case the pH will be driven down below the pH optimum of the enzyme, increasing the outward proton flux and decreasing the reaction rate/proton production until they equate. Thus the immobilized enzyme will always be forced to operate at a pH below its pH optimum regardless of the pH of the bulk phase.

If the reaction is initiated at a pH below the pH optimum then as protons are liberated the pH falls, reducing the reaction rate and the requirement for a proton gradient. Thus on the acid limb of the pH/activity profile proton accumulation in the matrix tends to be self-limiting. If experimental results are plotted as the ratio of the saturation rate obtained (V_{max}) to the saturation rate at the pH optimum (V'_{max}) versus pH_e, then the acid limb of the pH/activity profile of the immobilized enzyme will be very similar to the acid limb of the enzyme in dilute solution, while the alkaline limb may be broadened (Curve *A*, Figure 14) even to the extent of becoming a plateau (Curve *B*, Figure 14). In the extreme case outlined above where the enzyme is forced to operate below its pH optimum regardless of pH_e, the pH/activity profile will plateau before the pH optimum is reached (Curve *C*, Figure 14). The greater the restriction on proton diffusion and the higher the intrinsic activity of the enzyme, the more marked such perturbations will be. The converse pattern will be observed if the reaction utilizes protons.

A number of factors may complicate this pattern and these are discussed below.

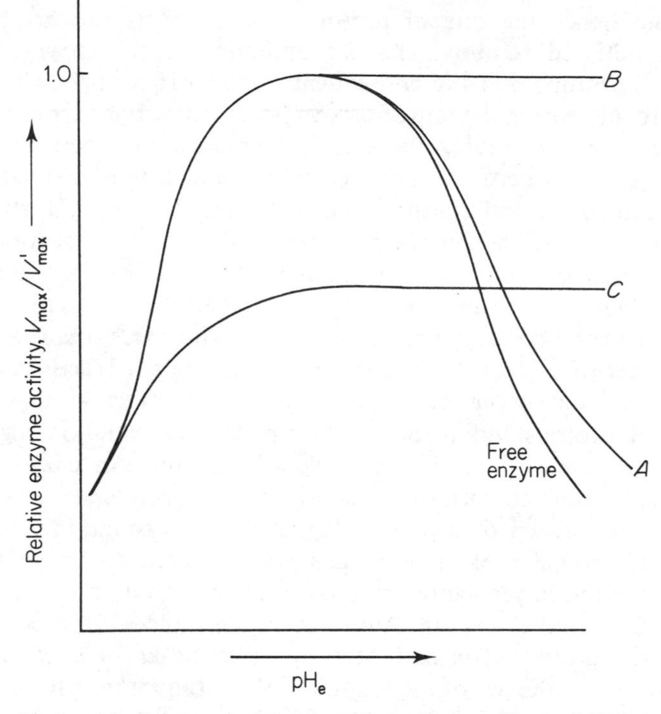

Figure 14. Schematic representation of the effect of H^+ion diffusion limitation on H^+ion liberating enzyme immobilized on to ionically neutral polymer. (*A*) mild, (*B*) moderate, and (*C*) severe diffusion limitation

3. The Presence of Buffers

Weak acids or bases act as buffers because of their ability to bind protons reversibly over a certain pH range. If an immobilized enzyme system such as the one just described is operating in the presence of a weak base (BH^+), then the base can act as a carrier for the protons released, facilitating their removal from the enzyme's microenvironment. In the low pH microenvironment of the enzyme, a greater proportion of the buffer will exist as BH^+ than in the higher pH of the bulk phase. BH^+ thus tends to diffuse into the bulk phase, where the higher pH causes it to dissociate releasing B which will diffuse down its concentration gradient back into the polymer matrix (see Figure 15). If the total buffer concentration is sufficiently high then protons will be rapidly removed from the microenvironment and little perturbation of the pH profile will be observed. When severe proton diffusion limitation is present, such that in the absence of any buffer the pH/activity profile becomes a plateau at values of pH below the pH optimum (Curve *C*, Figure 14), then the presence of a buffer may moderate the plateau, giving rise to an S-shaped acid limb to the pH profile. Figure 16 illustrates such an effect where it is assumed that the

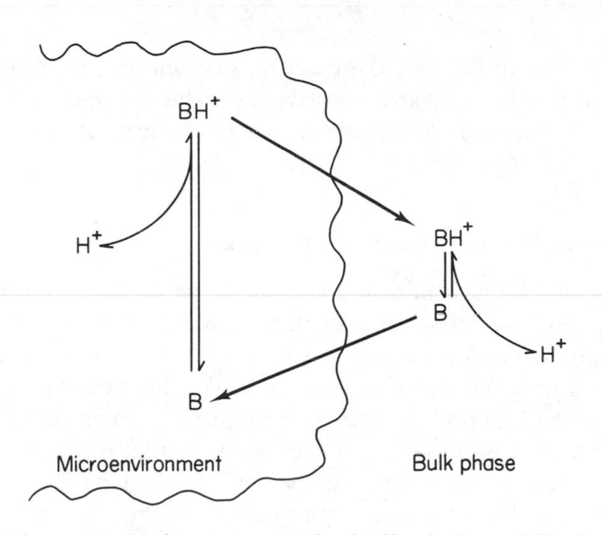

Figure 15. H^+ion transport by buffer in immobilized enzyme system

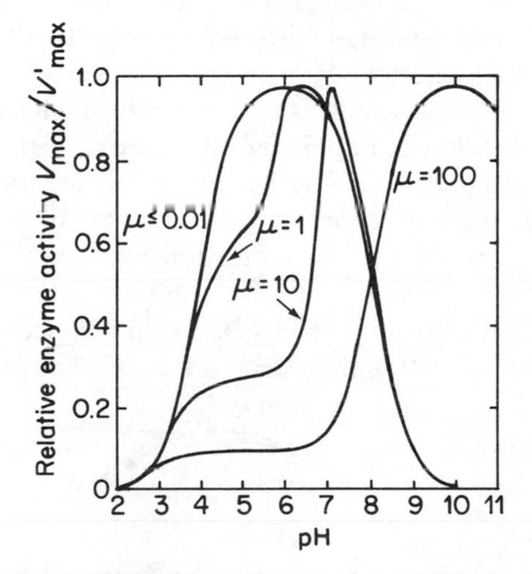

Figure 16. Effect of buffer on immobilized enzyme system demonstrating limitation of H^+ion diffusion. Buffer (10^{-3}mol 1^{-1}) pK_a assumed to be 8.0. μ represents the degree of diffusional resistance of the polymer matrix; thus curve $\mu \leq 0.01$ represents the free solution enzyme, curve $\mu = 100$ represents severe diffusional limitation. Reproduced with permission from Engasser and Horvath, *Biochim. Biophys. Acta*, **358**, 178 (1974)

pK_a of the buffer is 1.5 units higher than the pH optimum of the enzyme. At low pH$_e$ the buffer is fully protonated and hence cannot operate as a proton carrier. As the value of pH$_e$ is increased the enzyme activity rises and then the pH/activity profile becomes a plateau, until the pH$_e$ is raised to the point where the buffer begins to act as a carrier of protons from the microenvironment.

With a further rise in the pH_e the reaction rate will rise rapidly as the micro-environmental pH is increased towards the value of the enzyme's optimum pH. Ionizable substrates and products of the reaction, or even hydroxyl ions, may all act in this way as proton carriers.

4. Enzyme Loading and Substrate Limitation

The previous discussion would suggest that all perturbations in the pH profile are characterized by a broadening or displacement of the profile towards acid or alkaline values. However pH profiles have been observed where both limbs are broadened, thus the effective rate of the enzyme becomes less sensitive to pH changes when it is immobilized (see Figure 17). Such an effect can occur when neither proton partitioning nor diffusion limitation is present and is due to substrate diffusion limitation. An immobilized enzyme preparation that has a high enzyme loading (that is a large quantity of enzyme activity per unit of polymer) may be subject to substrate diffusion limitation. At the steady state, the rate of inward substrate diffusion at any point must equal its rate of removal by the enzyme. If the enzyme has a high intrinsic specific activity, the substrate concentration gradient through the particle will be steep and consequently the substrate may not penetrate to the centre of the immobilized enzyme particle. If some constraint is then applied which reduces the enzyme's activity (for example a change in pH) the substrate concentration gradient will become less steep thus allowing the substrate to penetrate further into the immobilized enzyme particle. More enzyme has now become available to the substrate; in effect the enzyme concentration has been raised. Two factors therefore work antagonistically on the reaction rate, the change in pH reducing the rate while the rise in effective enzyme concentration will tend to increase the rate, so moderating the effect of the pH change.

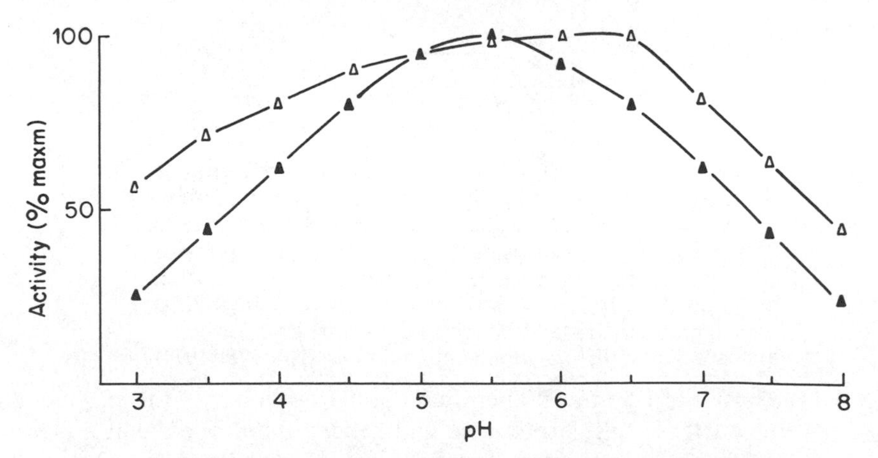

Figure 17. pH profiles of glucose oxidase, soluble (▲) and immobilized (△). Reproduced from Björek and Rosen (1976) by permission of John Wiley & Sons Inc

5. The pH Scale

When proton diffusion limitation is present the nature of the pH scale will cause a broadening of the alkaline limb of the pH/activity profile.

We have seen earlier (p.25) that a given rate of flux of protons will require an increase in pH gradient ($pH_i - pH_e$) as pH_i is raised. Consider an enzyme with a normal bell-shaped pH/activity profile immobilized in an ionically neutral proton diffusion limiting polymer (see Figure 18). Assume that at 0.5 V'_{max} the proton flux requires a (proton) gradient of 10^{-9}mol l^{-1}. At a value of pH_i of 7, pH_e will have to be 7.01 to satisfy this requirement and hence the acid limbs of the immobilized and free solution pH/activity profiles for the enzyme will not be very different. At a pH_i value of 8.9 however, the value of pH_e will have to be 9.6 in order to produce the same proton concentration gradient and proton flux. Thus the alkaline limb of the pH/activity profile of the immobilized enzyme will be broadened and displaced by 0.7 units at $0.5V'_{max}$. Such an effect will occur in addition to any autocatalytic effects of proton accumulation.

6. Combination of Partition and Diffusion Effects

When both proton partitioning and diffusion limitation are present the nature of the pH/activity profile may be difficult to predict. The number of possibilities fall into two categories, either partitioning and diffusion limitation may be working synergistically or they may be antagonistic. In the latter case it will be virtually impossible to predict the outcome.

It should be apparent that in all there are nine possible combinations of reaction type and support structure, proton liberating, proton utilizing, or 'neutral' reactions occurring in polyanionic, polycationic, or neutral polymer matrices. Of all these only the neutral support, 'neutral' reaction may show no perturbation in its response to pH variation and even this system will demonstrate a perturbed pH/activity profile if substrate limitation is present. The importance, therefore, of an understanding of the factors that may be involved in the response of an immobilized enzyme to pH cannot be underestimated.

It must be stressed that, in order to simplify the explanation of diffusion limitation, it has been implied that the microenvironment and the bulk phase exist as two homogeneous compartments separated by a discrete diffusion barrier. With the exception of microencapsulated enzymes, this is never the case. The microenvironment is a heterogeneous phase with respect to enzyme activity and solute flux rates, where, at a steady state, the rate of flux of a solute at a defined point will equal its rate of removal (production) by the reaction. The concentration gradients of solutes participating in the reaction within the polymer matrix are therefore non-linear (see Figure 7).

In conclusion it will be constructive to consider two systems that demonstrate the curious nature of the immobilized enzyme's dependence upon pH.

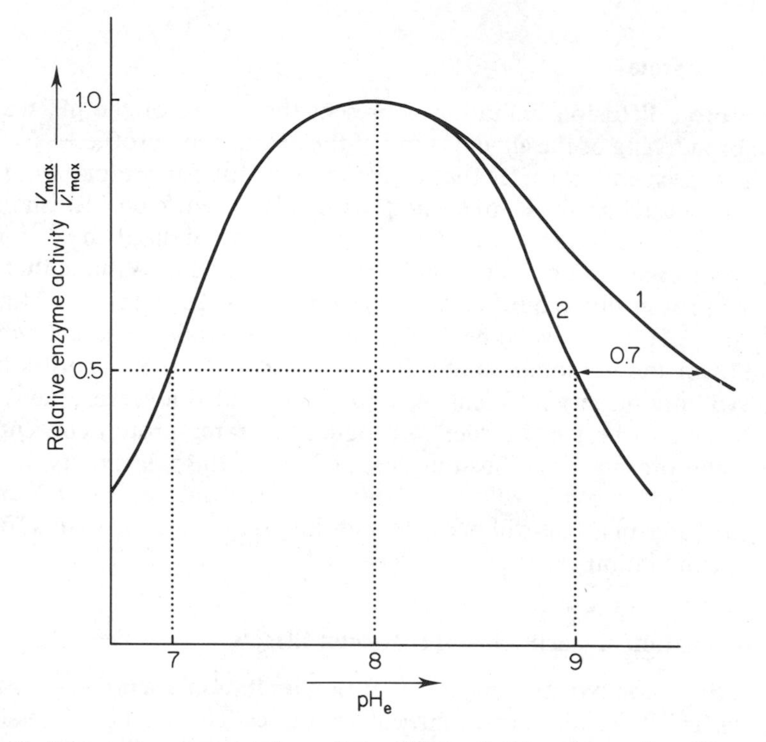

Figure 18. Schematic representation of the effect of pH scale on the pH profile of immobilized enzyme (1), compared to enzyme in dilute solution (2), for H^+ion liberating reaction

Ribonuclease cross-linked to a polyanion shows no change in its pH optimum when acting on RNA. This observation is not as surprising as it may seem at first sight, for RNA is itself a polyanion, thus even free solution ribonuclease is catalytically active in the close vicinity of a high concentration of negatively charged ions. In 1967 Tosa and his colleagues reported the pH dependence of papain immobilized within a polyanionic membrane. A small upward shift of the pH optimum was observed when benzoyl glutamate ethyl ester was used as a substrate. When the enzyme system was studied with benzoyl arginine ethyl ester, which exhibits a higher maximum catalytic rate than benzoyl glutamate ethyl ester, a much larger upward shift of the pH optimum was observed. The hydrolysis of both substrates releases H^+ ions, but clearly with its higher maximum rate of catalysis the hydrolysis of benzoyl arginine ethyl ester will form H^+ ions more rapidly than the hydrolysis of benzoyl glutamate ethyl ester, hence creating a greater perturbation of the immobilized papain's apparent pH dependence.

7. Electrostatic Fields

The observed shift of pH optimum of an enzyme upon immobilization to an ionic polymer may be explained in an alternative manner. Let us take the example of chymotrypsin, which is known to have on its surface 4 arginine, 14 lysine, 7 aspartate, and 5 glutamate residues. The molecule thus has a specific ionic atmosphere and a total of 6 positive charges at neutral pH. This positive electrostatic field stabilizes the basic form of the active site of the enzyme. Alteration of the surface charge of the molecule will perturb this stabilization effect and thus alter the pK_a controlling the pH dependence of the enzyme's activity. It can be shown experimentally that if the positive lysines on the surface are converted by reaction with succinic anhydride to (negative) carboxylates the characteristic pH/activity curve shifts 1 pH unit upwards (i.e. the observed pK_a of the ionization controlling catalytic activity shifts from 7.0 to 8.0). Conversely, rendering the surface of the chymotrypsin molecule more cationic by converting all the carboxylate groups to amines shifts the pH optimum 1 unit downwards.

This is exactly the pattern observed when α-chymotrypsin is immobilized on to polycationic or polyanionic supports (see Figure 12). Furthermore these shifts are destroyed by solutions of high ionic strength, both in the altered free solution enzyme and the unaltered immobilized enzyme.

IV. EFFECTIVE KINETIC PARAMETERS

We shall now turn to the question of the effective kinetic parameters of an immobilized enzyme. First, however, we must consider the meaning of the terms K_m and V_{max} which are usually applied to an enzyme in order to describe its catalytic characteristics. V_{max} and K_m are, by definition, the constants of a rate equation derived to describe enzyme activity in dilute solution. Under such conditions the equation relating the reaction velocity (V) and the substrate concentration (S) describes a rectangular hyperbole. When the reciprocals of V and S are plotted against each other the result is a straight line graph, the intercepts of which give the reciprocal values of K_m and V_{max}. So far so good. However a number of features may be present in an immobilized enzyme system that will produce curved reciprocal plots. In these circumstances it becomes impossible to determine K_m and V_{max} graphically and throws into question the wisdom of even using the terms K_m and V_{max} with their strict mathematical connotations. It is for this reason that it is better to redefine the kinetic parameters in the case of an immobilized enzyme. A number of alternative schemes have been developed, the earliest of which was to use K_m apparent, but for the purposes of this text we shall use the two constants K_v and V_s, whose similarity to the non-mathematical meaning of K_m and V_{max} will be obvious.

V_s we shall define as the highest possible velocity theoretically obtainable for a given system, i.e. when all of the enzyme is saturated with substrate. This parameter will therefore reflect the intrinsic characteristics of the immobilized enzyme but may be affected by diffusional constraints.

K_v can be defined as the substrate concentration that gives a reaction velocity of $V_s/2$. This parameter will reflect the effective characteristics of the enzyme and will depend upon both partitioning and diffusional effects.

Early reports on the effect of immobilization on an enzyme's kinetic parameters tended to investigate the relationship between S and V over a narrow range of substrate concentrations such that the reciprocal plots often appeared linear. Extrapolations of such plots led to the derivation of a parameter K_m (apparent). However the value of K_m (apparent) would vary depending upon the substrate concentration range used because, for example, the effect of substrate diffusion limitation will be greater at lower substrate concentrations and partitioning effects will be greater at low ionic strengths. Both of these factors will influence the apparent ease with which enzyme and substrate can associate and will thus affect the term K_m (apparent).

The remainder of this section is largely concerned with the factors that can cause non-linearity in reciprocal plots of V and S. It should be noted that although for the sake of clarity and convenience Lineweaver–Burk plots will be used, in practice Eadie–Hoftsee plots are much more useful for spotting non-Michaelis–Menten behaviour.

1. Partitioning Effects

Most of the early immobilization procedures relied upon the use of polyionic matrices. The reason for this was largely pragmatic, in that polyionic matrices are more hydrophilic than their non-ionic counterparts, thus they hydrate more readily.

It was rapidly discovered that the greater the degree of hydration the more active the immobilized enzyme preparation. However it soon became obvious that ionic interactions between the polymer and ionic substrates had a marked effect on the measured kinetic parameters of the enzyme.

The nature of the interaction between polyionic matrix and ionic solute is one of partitioning. Thus polycations will attract anions to their surface, polyanions will attract cations. For a given system the solute concentration at the surface of the polymer (X_i) will always be proportional to that in the bulk phase (X_o) regardless of the concentration of X_o, so long as the ionic strength remains constant (see p.23). Let us consider the theoretical case of an enzyme immobilized on to a polycation, acting on an anionic substrate S^-. S^- will be partitioned such that its concentration around the enzyme at the polymer surface (S_i) is greater than the bulk phase concentration (S_o). Figure 19 is a schematic Lineweaver–Burk plot for such a system, where an ionic partition coefficient of 2 has been used. The three obvious features of this plot are as follows. (1) The immobilized enzyme still obeys Michaelis–Menten kinetics. (2) The value of V_{max} remains unchanged. (3) The value of K_m appears to decrease by a factor of 2. In fact the Michaelis–Menten equation can be re-written to incorporate such partitioning effects (assuming that the ionic partition coefficient $p = S_i/S_o$) thus

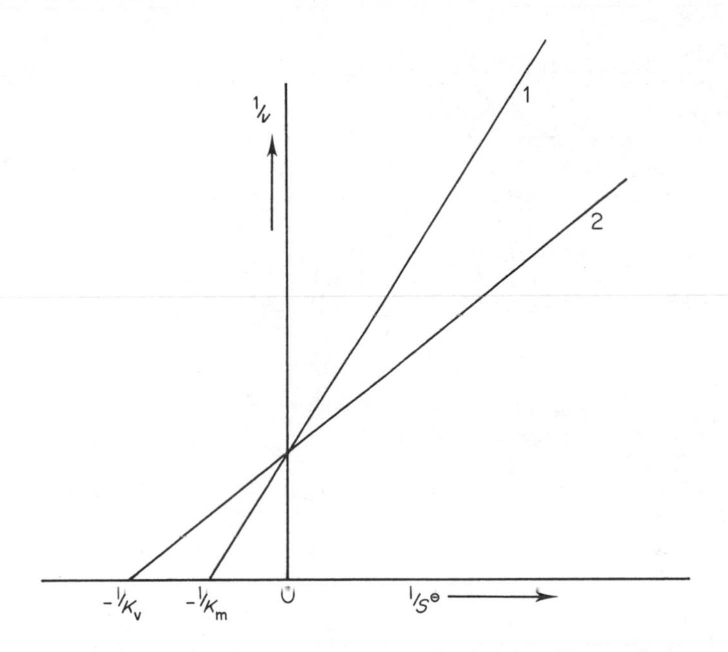

Figure 19. Schematic Lineweaver–Burk plot for dilute solution enzyme (1) and enzyme immobilized to polycationic polymer (2) acting on anionic substrate S^{-1}

$$V = \frac{V_{max} \cdot S_o \cdot p}{K_m + S_o \cdot p} \text{ and thus } K_v = \frac{K_m}{p}$$

Many reports have been presented showing that the apparent Michaelis constant is altered by up to one order of magnitude as a result of the interaction between ionic substrate and polyionic support. Table 2 presents just a few of these figures. Clearly the apparent K_i values for ionic inhibitors will be subject to the same form of constraint.

However the story does not end quite so simply. The ability of a polyionic matrix to partition ionic solutes will depend in part upon the ratio of number of solute ions present to matrix charge. Clearly, the greater the quantity of solute ions surrounding a particular matrix, the lower the magnitude of the partition effect on any one ion.

Thus the partition coefficient will decrease with increasing ionic strength of the solution and, as the partition coefficient decreases, so will the extent of the perturbation of the apparent K_m. This effect was demonstrated by Wharton and co-workers who immobilized bromelain on to carboxymethylcellulose and, using benzoyl-L-arginine ethyl ester (BAEE) as a substrate, studied the relationship between the apparent K_m and ionic strength (Figure 20). It can be seen from Figure 20 that the value of K_v rises with the ionic strength to reach a limiting value of about 50% of the K_m for the native bromelain, which is itself unaffected by ionic strength. This suggests that some factor other than

Table 2 Perturbation of observed K_m by immobilization

Enzyme	Support +	Charge	Substrate +	Charge	K_m(mol l^{-1})	Reference
ATP — creatine phosphotransferase	NONE	0	ATP	—	6.5×10^{-4}	1.
	p-Amino-benzoyl-cellulose	0	ATP	—	8.0×10^{-4}	
	Carboxy-methyl cellulose	—	ATP	—	7.0×10^{-3}	
Ficin	NONE	0	BAEE	+	2.0×10^{-2}	2.
	Carboxy-methyl cellulose	—	BAEE	+	2.0×10^{-3}	
Trypsin	NONE	0	BAA	+	6.8×10^{-3}	3.
	Ethylene-maleic anhydride	—	BAA	+	2.0×10^{-4}	

BAEE = beyzoyl arginine ethyl ester
BAA = benzoyl arginine amide

References 1. Hornby, W. E., Lilly, M. D., and Crook, E. M. (1968), see Bibliography.
2. Hornby, W. E. *et al.* (1966) *Biochem. J.*, **98**, 420.
3. Goldstein, L. (1964) *Biochemistry*, **3**, 1913.

ionic partitioning is also contributing to the perturbation of the K_m. Theoretical calculations also showed that the value of K_v reached half the limiting value when the ionic strength of the bulk solution was half the value of the concentration of the carboxylate groups of the carboxymethylcellulose measured in their own volume.

It should be apparent that the (ionic) substrate will contribute to the ionic strength, and may itself mask the partitioning effect of the polymer, particularly at high concentrations. In these circumstances a reciprocal velocity/substrate concentration plot will be sigmoidal not linear.

2. Diffusion Limitation

Wishful thinkers might hope that, for a given system, V_s will reflect either the intrinsic characteristics or the effective characteristics of the immobilized enzyme. Unfortunately for the tidy-minded temperature has its effect, for while an immobilized enzyme system may be effectively controlled by kinetic constraints at low temperatures, as the temperature is raised substrate diffusion becomes the rate-limiting factor. This occurs essentially because temperature has a more marked effect on enzyme reaction rates than it does on rates of diffusion and the effect may be conveniently illustrated by a schematic

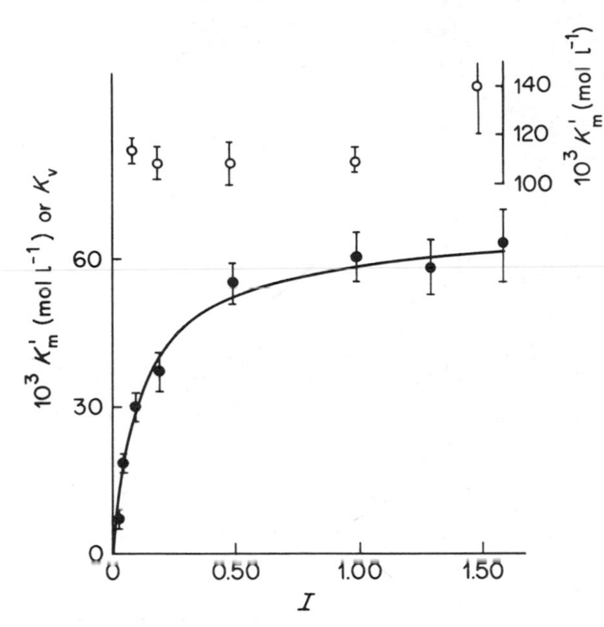

Figure 20. Effect of ionic strength on apparent Michaelis constant $K_{m'}$ of bromelain in dilute solution (○) and immobilized to carboxymethylcellulose (●), acting on benzoyl arginine ethyl ester. Reproduced with permission from Wharton *et al.*, *Eur. J. Biochem.*, **6**, 572 (1968)

Arrhenius plot of an immobilized enzyme (Figure 21). The low temperature section of the plot is kinetically controlled and reflects the true activation energy. In the mid-temperature section of the plot the rate is limited by diffusion constraints within the matrix. The limiting factor in the high temperature section of the plot is the rate of diffusion of substrate to the matrix which, compared with the enzyme reaction rate, is so insensitive to temperature changes the slope of the Arrhenius plot is effectively zero. Thus whether or not V_s reflects any diffusional constraints will depend upon the temperature. By constructing an Arrhenius plot for a particular enzyme system, the activation energy of the immobilized enzyme can be compared to the dilute solution enzyme in the absence of diffusional constraints. This may provide information on the effect of the immobilization process upon the enzyme's intrinsic properties (p.16) or may be used in the comparison between an enzyme immobilized or in its natural intracellular environment (p.124).

In basic texts on enzymology it is often stated that the parameter K_m is an inverse measure of the affinity of the enzyme for the substrate. When an enzyme is placed in a heterogeneous environment the physical factors present may act either to impede or to facilitate the approach of the substrate to the enzyme. Thus the apparent affinity of the enzyme for the substrate may either

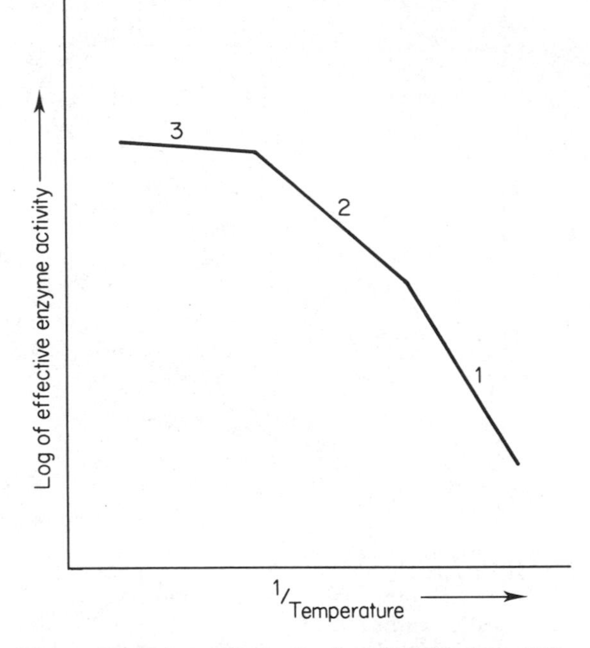

Figure 21. Schematic Arrhenius plot for immobilized enzyme showing three regions. (1) Reaction under kinetic control. (2) Reaction under internal diffusion control. (3) Reaction under external diffusion control

be decreased or increased and thus K_v may be smaller or larger than the intrinsic K_m.

Mathematically K_m for a simple unisubstrate enzyme reaction may be defined by a combination of the rate constants for the individual steps in the reaction. At its simplest it is represented by equation.

$$E + S_i \underset{K_{-1}}{\overset{K_1}{\rightleftharpoons}} ES_i \xrightarrow{K_2} E + P$$

$$\text{Hence } K_m = \frac{K_{-1} + K_2}{K_1}$$

For all enzyme reactions in dilute solution under standard conditions K_1 represents the speed with which substrate and enzyme may diffuse together and collide and is essentially the same for all enzyme–substrate collisions. If we assume that K_{-1} and K_2 are constant for a given system, then it will be apparent that any factor that affects the ease with which enzyme and substrate may come into contact will affect the apparent value of K_m, not by altering K_1

but by adding a new term to the equation,

$$S_o \xrightarrow{h_s} S_i + E \rightleftharpoons ES \longrightarrow E + P$$

Where h_s is a constant describing the rate of diffusion of the substrate to the enzyme. If

$$h_s \gg \frac{K_1 E_T}{(K_{-1} + K_2)K_2} = \frac{(V_{max})}{K_m}$$

then the reaction will be kinetically controlled and the effective first order rate constant $K' = K_m/V_{max}$. Conversely when $h_s \ll V_{max}/K_m$ the reaction will be diffusion controlled ($K' = h_s$) and the effective reaction velocity $V' = V_{(diff)}$ the diffusion velocity. A schematic representation of the relationship is given in Figure 22. Clearly, at sufficiently high substrate concentrations $V' = V_{max}$ the intrinsic maximum velocity of the reaction, but the substrate concentration that gives a value of $V_{max}/2$ (K_v) will always be greater than the true intrinsic K_m.

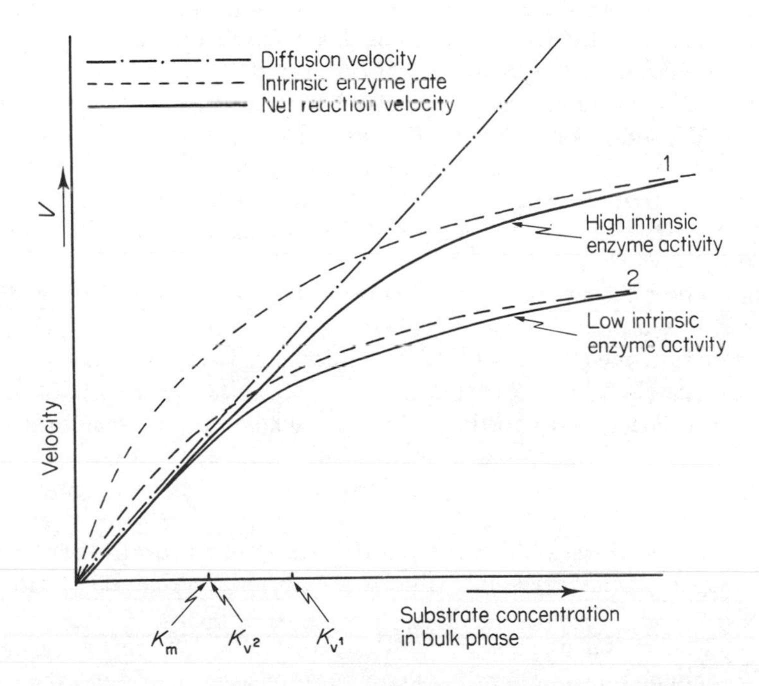

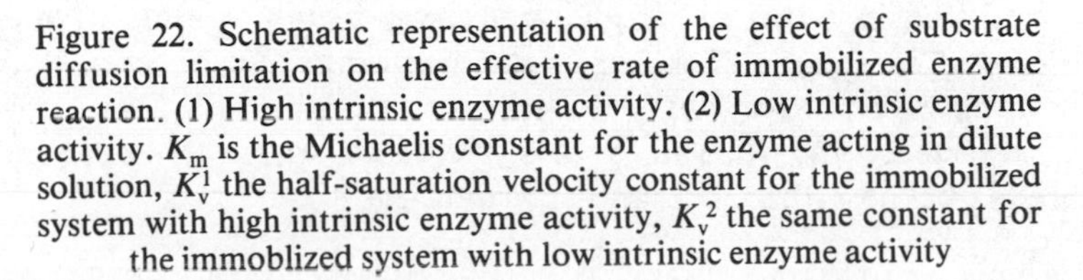

Figure 22. Schematic representation of the effect of substrate diffusion limitation on the effective rate of immobilized enzyme reaction. (1) High intrinsic enzyme activity. (2) Low intrinsic enzyme activity. K_m is the Michaelis constant for the enzyme acting in dilute solution, K_v^1 the half-saturation velocity constant for the immobilized system with high intrinsic enzyme activity, K_v^2 the same constant for the immoblized system with low intrinsic enzyme activity

The above consideration applies to the theoretical case where the enzyme is attached to an impervious polymer surface and the limit on substrate diffusion is through the unstirred layer between the bulk solution and the surface of the polymer. This type of diffusion limitation is known as external diffusion limitation and is affected by temperature and stirring rate.

The other form of diffusion limitation to be considered is the diffusion of substrate within the matrix of an immobilized enzyme particle, internal diffusion limitation. Here the picture is complicated mathematically because diffusion and catalysis are occurring concurrently rather than sequentially as is the case for external diffusion limitation. The qualitative effect on K_v will be similar to that for external diffusion limitation, that is $K_v >$ intrinsic K_m. In theory at least V_s will tend towards the intrinsic V_{max} if the substrate concentration is raised to a sufficiently high level. In practice, however, particularly in the case of a system with a high catalytic rate and/or severe internal diffusion limitation, the substrate concentration profile within the polymer matrix may be so steep that the substrate may not penetrate right to the centre of the particle. Thus some of the immobilized enzyme at the centre of the particle will remain inactive and V_s will appear to be less than the intrinsic V_{max}.

To summarize, diffusional limitations on the substrate's access to the immobilized enzyme will invariably give rise to a value of K_v greater than the intrinsic K_m, while in practice V_s may appear to be less than the intrinsic V_{max} where internal diffusion is a limiting factor. The size of these perturbations will depend upon the extent of substrate diffusion limitation in relation to the catalytic activity of the enzyme, the higher the intrinsic catalytic activity the greater the perturbation.

Many observations have been made of the effect of substrate diffusion limitation. Goldman and co-workers (1968a) constructed three papain–cellulose nitrate membranes of differing thickness. Using a relatively ineffective substrate acetyl-L-glutamic acid diamide (AGDA) the value of V_s obtained for each membrane was proportional to the thickness of the membrane. When they measured V_s using the effective substrate benzoyl-L-arginine amide (BAA), the values of V_s they obtained were much the same for each membrane.

A similar result was obtained in another study by the same workers with an alkaline phosphatase–cellulose nitrate membrane, acting on *p*-nitrophenyl phosphate.

Three thicknesses of membrane (1.6, 2.6, 8.8 $\times$ 10^{-4}cm) were prepared, each containing the same density of enzyme molecules. Curiously the V_s values when expressed in terms of enzyme activity per unit volume of membrane increased with increasing thickness of the membrane (23, 32, and 34 μmol s^{-1} cm^{-3}) an observation for which the experimenters provided no explanation. Not unexpectedly the K_v values also increased markedly with increasing membrane thickness (0.85, 2.9, and 12.0 mmol l^{-1}).

3. The Effect of pH

Thus far we have seen how diffusional restrictions and partitioning by the polymer matrix may affect an immobilized enzyme's response to changes in substrate concentrations. It is, however, quite possible to obtain non-linear reciprocal plots even when the substrate is homogeneously distributed throughout both the bulk and immobilized phases.

Let us consider the hypothetical case of an enzyme catalysing an acid-releasing reaction which in dilute solution exhibits a bell-shaped pH/activity curve. Such an enzyme is immobilized in a matrix which will restrict the free diffusion of H^+ ions at neutral or alkaline pH but not, because of its vastly greater concentration, substrate. If the immobilized enzyme is studied in essentially buffer-free conditions considerable perturbations of the pH/activity profile may be observed. The extent of such perturbations, as we have seen, will depend in part upon the intrinsic catalytic activity of the enzyme; the greater the rate of reaction the greater the perturbation. One factor which will influence the reaction rate is the substrate concentration. Thus with small substrate concentrations the reaction rate and rate of H^+ liberation will be low and the internal pH of the matrix will be little different from that of the bulk solution. If the substrate concentration is relatively high then the reaction rate will also be high and thus the internal matrix pH may be markedly different from the pH of the bulk phase. Herein lies the problem, the change in the pH of the enzyme's microenvironment brought about by the change in substrate concentration will obviously affect the rate of the reaction in addition to the increase in rate brought about by the raised substrate concentration. If the bulk phase pH is greater than the pH optimum of the enzyme these two factors will act synergistically (positive cooperation) or if the bulk phase pH is less than the pH optimum they will act antagonistically (negative cooperation). Figure 23 is a schematic representation of the pH/activity profile of an enzyme in dilute solution and immobilized in an appropriate manner. Consider first the effect of studying the enzyme at a bulk phase pH of 8. At low substrate concentrations the internal matrix pH will be approximately 8 and the immobilized enzyme's activity will be similar to that of the free solution enzyme (point *A*, Figure 24). At an intermediate substrate concentration (point *B*, Figure 24) the reaction will proceed initially at a rate characteristic of that substrate concentration at pH 8. However the increased rate of production of H^+ ions will lead to a reduction in the internal pH of the matrix which will in turn lead to an increase in the reaction rate. A steady state will be reached where the internal matrix pH is less than 8 (for example 7.5) and the reaction rate is consequently greater than might be predicted from the rise in substrate concentration alone. Thus the measured reaction rate is greater than that of the free solution enzyme at pH 8.

A reciprocal plot in such a system will thus be curved. The shape of this curve will depend upon the complex interrelationship between the enzyme's

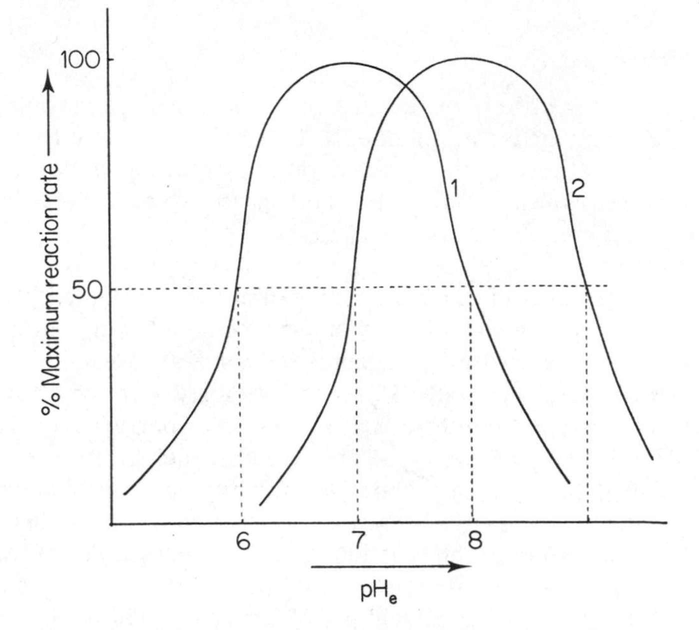

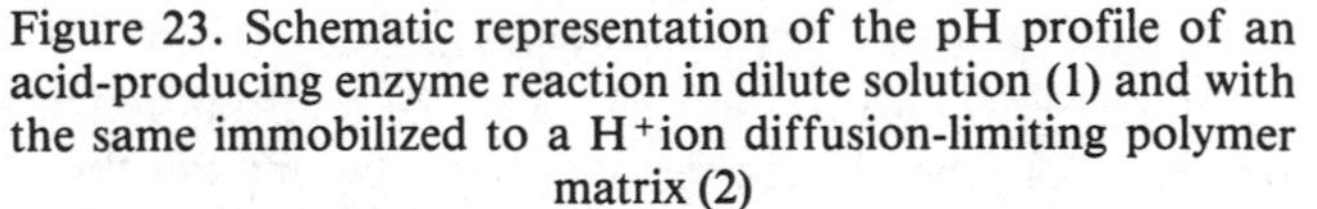

Figure 23. Schematic representation of the pH profile of an acid-producing enzyme reaction in dilute solution (1) and with the same immobilized to a H^+ion diffusion-limiting polymer matrix (2)

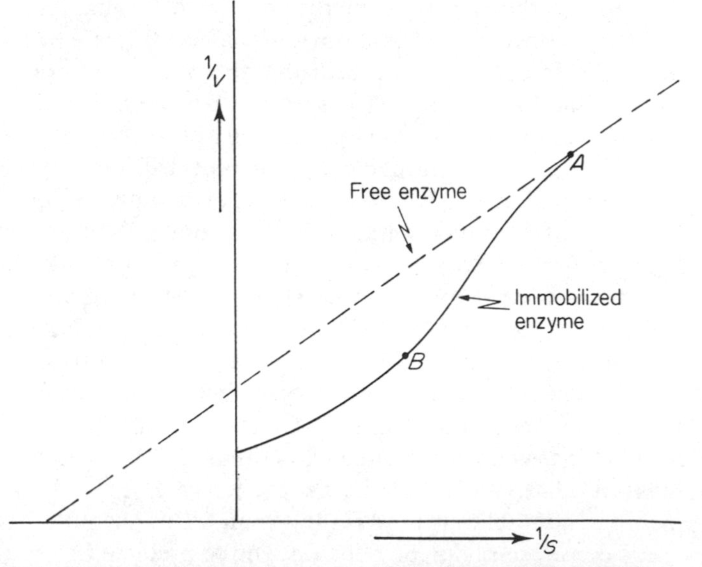

Figure 24. Reciprocal Lineweaver–Burk plot of acid-producing enzyme reaction in dilute solution (broken line) and for the enzyme immobilized to a H^+ion diffusion-limiting polymer matrix (solid line) in the total absence of substrate and product diffusion limitation. The system is assumed to be acting above the pH optimum of the enzyme (see Figure 23)

response to pH changes and the extent of H^+ion diffusion limitation by the matrix. In theory at least it should be an *S*-shaped curve (see Figure 24) but in practice it may appear either convex or concave depending upon the particular system. With severe H^+ion diffusion limitation and a steep pH/activity profile the curve will become predominantly convex towards the $1/S$ axis.

The converse will apply if the enzyme reaction takes place at a bulk phase pH below the pH optimum, e.g. pH 6 in this hypothetical case. Again at low substrate concentrations the reaction rate will be small, H^+ion production low, and the internal matrix pH will be more or less the same as that of the bulk phase. If the substrate concentration is raised, the rate of H^+ion production increases and thus the internal pH falls. This will have the effect of reducing the reaction rate below the value that might be predicted from knowledge of the substrate concentration. The consequent effect of this will be an S-curved reciprocal plot which in practice will appear either predominantly convex or concave.

In all of the above we assume that K_m is pH invariant. If we now relinquish this assumption it will be seen that the situation is even more complex. If K_m changes with pH it may either alter in the same direction as V_{max} (e.g. K_m and V_{max} both increase with increasing pH), or in the opposite direction. Where the change in V_{max} and K_m are in the same direction then any potential curvature in the reciprocal plot will be moderated and where the changes are in opposite directions the curvature will be exaggerated. In the ideal case where V_{max} and K_m both change with the same rate and direction with pH, a straight line reciprocal plot will result as the gradient of such a plot is the ratio K_m/V_{max}.

So much for the theory. The astute reader will probably comment on the inadvisability of attempting to plot reciprocal or other kinetic plots of immobilized enzymes. While there are obvious implications in this respect, the real importance of these effects is twofold. First, to the biochemical engineer constructing an immobilized enzyme reaction system it will be apparent that, if insufficient buffering capacity is available, then attempts to increase the rate of the reaction by increasing the substrate concentration may be ineffective. Such problems are discussed at greater length in Chapter 3, Section IV. Second, to the academician attempting to understand the complexities of control within the cell, it is apparent that the kinetic effects produced in the hypothetical system described above are very similar to the positive and negative cooperation of allosteric enzymes. Cooperative effects in an enzyme's response to changes in substrate concentration are important because they allow both great change in reaction rate over a small substrate concentration range and small changes in reaction rate over other, very large, substrate concentration ranges. Thus over an appropriate substrate concentration range an allosteric enzyme, unlike a Michaelis–Menten enzyme, can be made either to respond very markedly to fluctuations in substrate concentration or to be virtually independent of such fluctuations. However, as we have seen, it is possible to bestow these same attributes on non-allosteric enzymes by providing an appropriate microenvironment and, whereas one particular allosteric enzyme may exhibit either positive or negative cooperation, the immobilized enzyme may show both positive and negative cooperation depending upon the pH surrounding its microenvironment. The local control of H^+ion concentration and fluxes within a cell may therefore be of prime importance to metabolic control.

We have considered throughout this section the effect of H^+ions on the kinetic manifestation of an immobilized enzyme. Such effects may be treated in a similar manner to the effects of enzyme activators or inhibitors and where the H^+ions are a product of the reaction they may be treated as a special case of product inhibition (or activation). This leads us then to the consideration of the effects of chemical inhibitors on immobilized enzymes.

4. Inhibition

For the purposes of the present discussion we may divide enzyme inhibition into four categories, inhibition by low molecular weight solutes, high molecular weight solutes, products of the reaction, and substrate inhibition. Each of these will be dealt with separately.

a. Low Molecular Weight Inhibitors

Low molecular weight inhibitors acting on an immobilized enzyme may be subject to exactly the same constraints as the substrate of that enzyme, that is they may be partitioned by the polymer matrix and/or their free diffusion restricted. Generally speaking, and with the exception of product inhibition, the restriction of the free diffusion of an inhibitor molecule does not affect the consideration of the steady-state rate of an immobilized enzyme reaction. The reason for this is quite straightforward, i.e. in the absence of partitioning effects, any non-reacting solute will be homogeneously distributed throughout the bulk and enzyme phases once the system has equilibrated.

The relevant factors we must consider then are partitioning of the inhibitor in the presence or absence of substrate partitioning and/or diffusion limitation.

The simplest of these possibilities is the case of inhibitor partitioning in the absence of any effect of the matrix on the substrate. The inhibitor concentration in the microenvironment will be either higher or lower than that of the macroenvironment, depending upon the ionic characteristics of inhibitor and polymer. The effect on the pattern of inhibition observed will be to alter the degree of inhibition at a given bulk phase inhibitor concentration by a factor related to the partition coefficient of the inhibitor.

Thus when inhibitor and polymer are of the same charge the degree of inhibition will be reduced, when inhibitor and polymer are oppositely charged the degree of inhibition will be increased. When substrate partitioning also occurs then the degree of inhibition will be greatest when inhibitor and substrate and inhibitor and polymer have opposite charges; the degree of inhibition will be least when inhibitor and substrate and substrate and polymer have opposite charges. In the case where a competitive inhibitor and substrate have the same charge and the same micro/macroenvironment partition coefficient, the degree of inhibition will be identical to that exhibited by the free solution enzyme. These considerations apply equally to all kinetic types of inhibition including product inhibition, and in all cases the enzyme system will apparently obey Michaelis–Menten kinetics.

When substrate diffusion is also restricted the degree of inhibition will depend additionally on the extent of diffusion limitation. The qualitative logic of the argument is identical to that for the combined effect of substrate diffusion limitation and enzyme inactivation (p.19). It can be observed both experimentally and theoretically that when severe substrate diffusion limitation is present an immobilized enzyme may show little or no response to the addition of an inhibitor. The reduction in the inherent activity of the enzyme in such circumstances has no effect on the observed rate of reaction because the rate is primarily determined by the rate of diffusion of substrate into the enzyme phase. At sufficiently high substrate concentrations the reaction becomes kinetically controlled and then the presence of an inhibitor may be apparent (see Figure 25). At the other extreme, where substrate

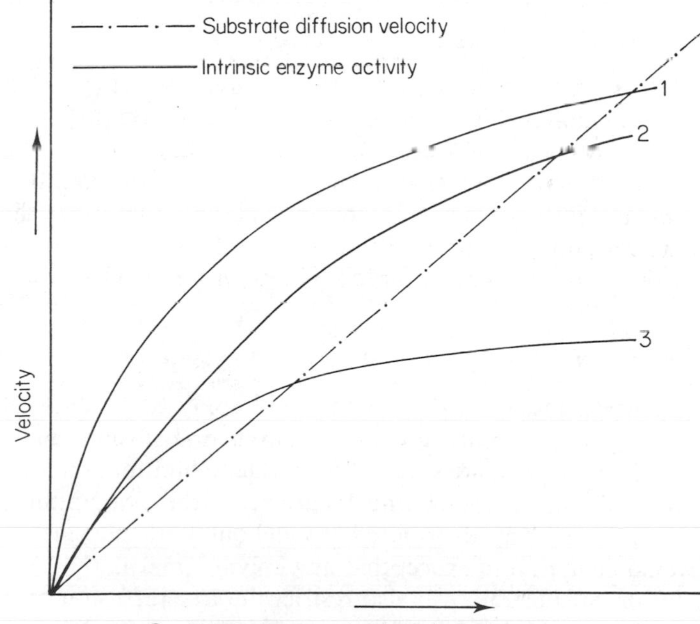

Figure 25. Schematic representation of the interaction of substrate diffusion limitation and chemical inhibition for an immobilized enzyme. Solid lines represent intrinsic enzyme activity of immobilized enzyme, uninhibited (1), subjected to competitive inhibition (2), and subjected to non-competitive inhibition (3). Clearly competitive inhibition may not be apparent if substrate diffusion limitation is severe

diffusion limitation is negligible, the effect of the inhibitor will be identical to that of the enzyme in free solution. In between these two extremes the degree of inhibition produced by an inhibitor will become progressively less as the extent of substrate diffusion limitation increases.

In these intermediate cases the reduction in enzyme activity caused by the inhibitor reduces the relative diffusion limitation of the substrate, thus the chemical inhibition moderates the diffusional inhibition and the combined effect is less than their sum. This may be accounted for by an alternative rationale. An immobilized enzyme particle subject to substrate diffusion limitation will contain a concentration gradient of substrate. The slope of this concentration gradient will depend upon the ability of the substrate to diffuse within the matrix and upon the inherent activity of the enzyme. The greater the inherent activity of the enzyme, the less far the substrate will be able to penetrate into the microenvironment. An inhibitor which reduces the inherent enzyme activity will therefore allow the substrate to penetrate further into the polymer particle and increases the total number of enzyme molecules available to the substrate. Thus the reduction in inherent enzyme activity is offset to some extent by a rise in the effective enzyme concentration.

b. High Molecular Weight Inhibitors

Certain inhibitors may be of a similar molecular size to the enzyme itself. The example of trypsin inhibited by a peptide of pancreatic origin (MW 6,500) or a peptide from the soya bean (MW 21,000) is given on p.20).

The gross steric restrictions that may be placed on the ability of such inhibitors to come into contact with the enzyme and the effect that these restrictions produce on the pattern of inhibition, will occur in addition to any of the effects that may be observed for low molecular weight inhibitors.

c. Product Inhibition

The general considerations discussed above will apply to product inhibition when partitioning effects only are present. When diffusion limitation is present within the polymer matrix then, for product inhibition, unlike all other inhibition so far discussed, we can no longer make the assumption that the inhibitor is homogeneously distributed throughout both enzyme and bulk phases. It would be logical to expect that any polymer matrix that can restrict the free diffusion of substrate will also restrict the free diffusion of product, especially since the concentration of product is almost invariably far smaller than that of the substrate.

Let us now consider the implications of product inhibition in the presence of diffusion limitation. The first, and perhaps most obvious, implication is that the microenvironmental concentration of product must be greater than that of the bulk phase, and it is possible, therefore, for an immobilized enzyme to be subjected to product inhibition when the bulk phase product concentration appears insignificant. Second, if the immobilized enzyme system is one where diffusion within, rather than to, the polymer matrix is the limiting factor then, while the substrate concentration will decrease towards the centre of the polymer particle, product concentration will increase. Thus the degree of chemical inhibition will be greatest where the substrate concentration (and inherent kinetic rate of reaction) is least and diffusion limitation greatest. As we have seen the expected degree of inhibition is decreased as the degree of diffusion limitation is increased, therefore in this case product inhibition will be less marked than expected. For the purposes of qualitative analysis we shall take a slightly simpler system where the enzyme is bound to an impervious surface, thus substrate and product concentrations around the enzyme are uniform (but differ from those of the bulk solution). At steady state the product must diffuse away from the enzyme at the same rate that the substrate approaches it. When, for a given system with fixed macroenvironmental substrate concentration, the product concentration is raised, the enzyme rate and thus the effect of diffusional limitations are lowered so that, at sufficiently high product concentrations substrate depletion and product accumulation vanish and the inhibition exhibited will be entirely due to chemical interaction of product and enzyme. If the product concentration is lowered then the inherent enzymic rate will increase and diffusional resistances again become apparent; substrate will be depleted and product will accumulate around the enzyme, both events increasing the total degree of inhibition experience by the enzyme (Figure 26).

Whereas the pattern described above will occur with either competitive or non-competitive product inhibition, variation of the substrate concentration, in the presence of a fixed product concentration, will provide differing results depending upon the nature of the inhibition. Where the product is a competitive inhibitor, at sufficiently high substrate concentrations the inherent V_{max} value of the enzyme may be obtained, that is a point is reached where both diffusional limitation and product inhibition are both overcome. In the case of non-competitive product inhibition the effective saturation velocity V_s is always less than the inherent maximum velocity, because increasing the substrate concentration merely increases the enzyme reaction rate, making product accumulation more likely. Furthermore, the more effective the polymer is as a barrier to free diffusion the lower will be the value of V_s (Figure 27).

d. Substrate Inhibition

The normal pattern of substrate inhibition of an enzyme acting in free solution is shown schematically in Figure 28. When considering any effects of enzyme immobilization on the perceived pattern of substrate inhibition it is important to distinguish between partitioning and diffusional effects. In the former the considerations are relatively straightforward, the latter may give rise to multiple steady states and therefore we shall deal with the two situations separately.

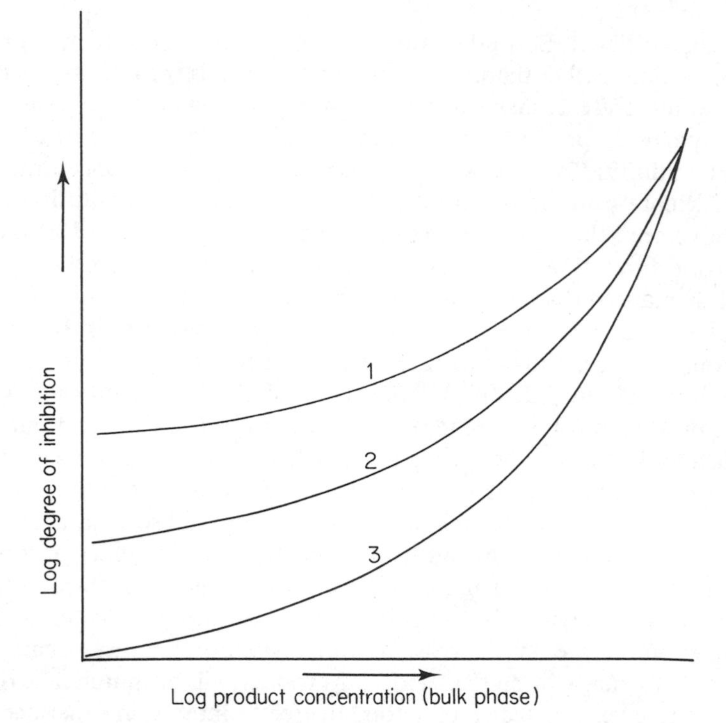

Figure 26. Schematic representation of the effect of product concentration on degree of inhibition exhibited by immobilized enzyme system subject to product inhibition, (1) in the presence of both substrate and product diffusional limitation, (2) with substrate diffusion limitation, and (3) in the absence of diffusional restriction. Bulk phase substrate concentration remains at a fixed value throughout

As we have already observed polyionic matrices will cause a partitioning of any ionic solutes between the matrix surface and the bulk phase. In the general case of an enzyme subject to inhibition by its anionic substrate and immobilized on to a polycationic support, the substrate concentration around the enzyme will be greater than that in the bulk phase. Thus any inhibitory effects the substrate might demonstrate will become apparent at a lower bulk phase concentration (Figure 28). The converse will obviously apply when support matrix and substrate bear the same ionic charge. So long as the ionic strength remains constant the observed effect will be an apparent shift of the optimum substrate concentration to higher or lower values as if the substrate concentration axis has been stretched or compressed by a value corresponding to the partition coefficient. If no attempt is made to maintain a constant ionic strength then the rise in ionic strength that accompanies an increase in (ionic) substrate concentration will reduce the partitioning effect. Thus at high substrate concentrations there will be no effective partitioning and the onset of

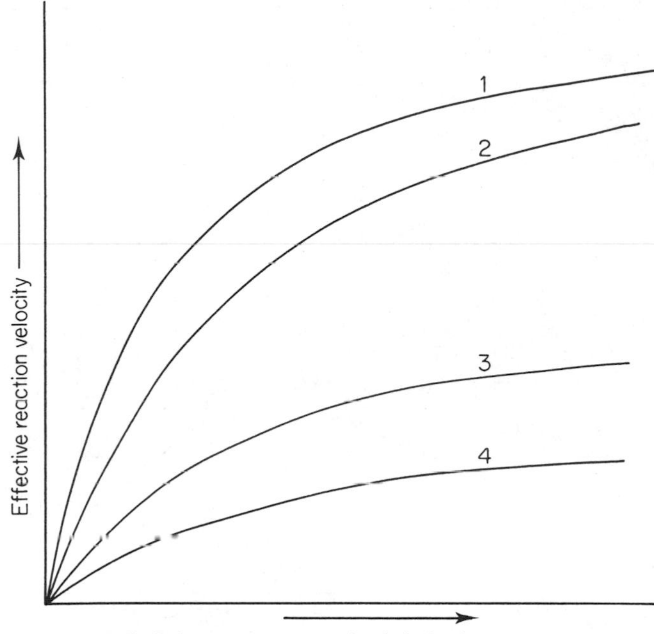

Figure 27. Schematic representation of the effect of substrate concentration on reaction velocity of immobilized enzyme reaction in the presence of a fixed concentration of product inhibitor with diffusional limitation. (1) Intrinsic uninhibited enzyme activity. (2) Competitive product inhibition. (3) Non-competitive product inhibition with slight accumulation. (4) Non-competitive inhibition with severe product accumulation

substrate inhibition will occur at the same bulk phase substrate concentration for both free and immobilized enzyme.

Let us now turn to the case of an immoblized enzyme subject to both substrate inhibition and substrate diffusion limitation. The observed dependence of enzymic activity on the bulk phase substrate concentration will be governed by two interrelated factors, the degree of substrate diffusion limitation and the bulk solution substrate concentration. In this context it must be remembered that at low substrate concentrations the reaction will tend to be diffusion controlled while at high substrate concentrations the reaction will be kinetically controlled (see Figure 22). The higher the intrinsic activity of the enzyme the more important the effect of diffusion limitation; conversely an enzyme with a low intrinsic activity will exhibit relatively little substrate diffusion limitation. When these considerations are combined with the effect of substrate inhibition, it should be apparent that the reduction in enzyme activity brought about by a rise in substrate concentration may suddenly switch the reaction from a diffusion controlled state to a kinetically controlled state.

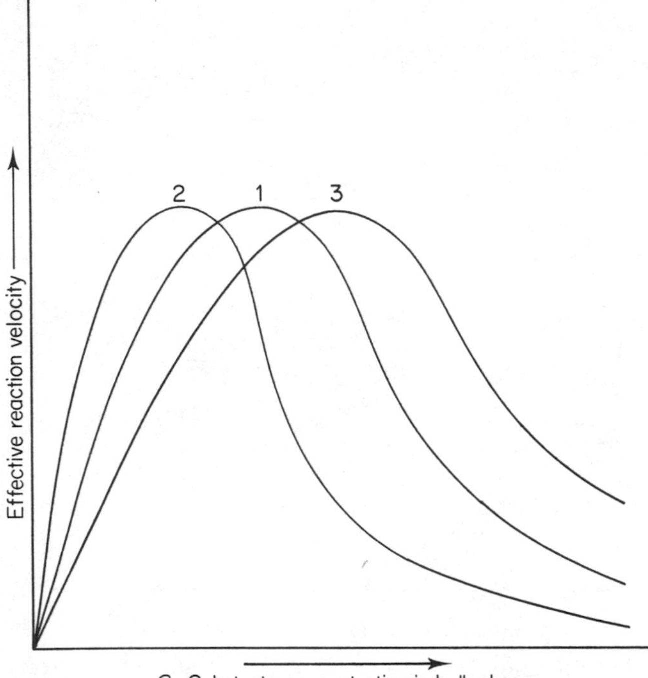

Figure 28. Schematic representation of the effect of concentration of anionic substrate on reaction velocity of substrate-inhibited enzyme, (1) in dilute solution, (2) immobilized on polycation, and (3) immobilized on polyanion

Substrate diffusion limitation effectively reduces the substrate concentration in the enzyme phase (S_i) below that of the bulk phase (S_0), so that a higher value of S_0 is necessary before the onset of substrate inhibition than is the case with the enzyme in free solution (Figure 29, curve 2).

For a fixed bulk phase substrate concentration (S_0) the rate of transport of substrate to the matrix surface will depend upon the substrate transport coefficient h_s, and the substrate concentration at the surface of the matrix S_i. Figure 30 shows this relationship and is in effect a restatement of Fick's law of diffusion. An immobilized enzyme reaction reaches a steady state when the rate of transport of substrate to the surface equals the rate of reaction (i.e. rate of substrate removal). If we combine Figures 29 and 30 (Figure 31) the effective steady-state rate of the enzyme reaction will be described by the point(s) of intersection of the curve of the intrinsic kinetic rate of the enzyme with the curve of the rate of transport of substrate to the enzyme. It can be seen from Figure 31 that, when substrate diffusion limitation is minimal (i.e. h_s is high) or when it is maximal (i.e. h_s is very low) only one intersection occurs

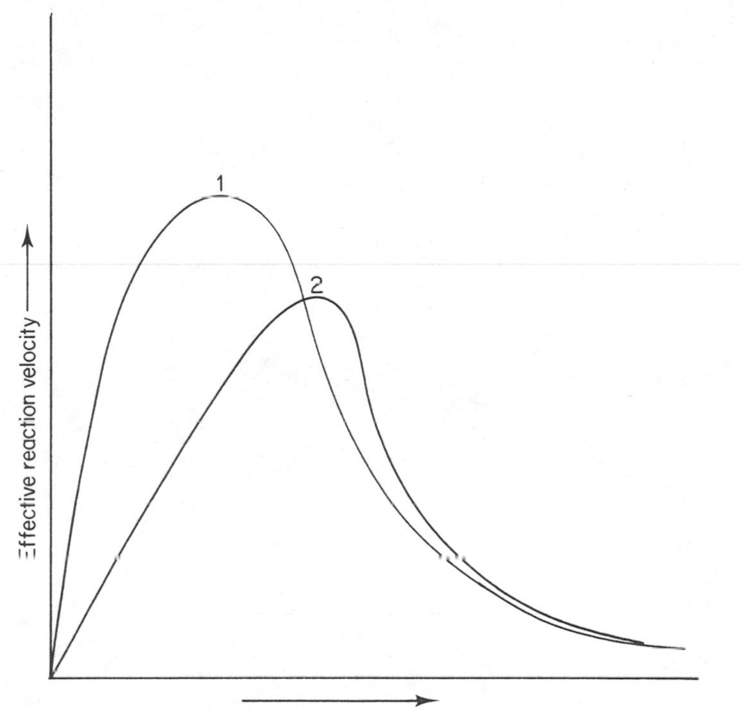

Figure 29. Schematic representation of the effect of bulk phase substrate concentration (S_0) on substrate-inhibited enzyme, (1) in dilute solution and (2) immobilized to a polymer matrix exhibiting substrate diffusion limitation

and therefore only one steady-state rate is permissible. At intermediate values of h_s three intersections are possible and thus three different steady-state rates (*A*, *B*, and *C* in the Figure 31) are permissible. An exactly similar situation occurs if we consider the value of h_s to be fixed and S_0 is varied (Figure 32). If we now plot S_0 against the effective enzymic rate, at high and low values of S_0 only one steady-state rate is possible, but at intermediate values of S_0 the three rates *A*, *B*, and *C* are possible and the result is a hysteresis curve (Figure 33).

Although theoretical predictions give rise to such a curve, in practice a hysteresis loop (Figure 34) is formed as the centre line section is not allowable. The existence of a hysteresis loop means that the actual enzymic rate will depend not only on the bulk phase substrate concentration but on its direction of change. Thus moving from a low bulk phase substrate concentration to higher values will increase the reaction rate, but because of the presence of substrate diffusion limitation the optimum value of S_0 is shifted upwards. As S_0 is further increased a point is reached where the system switches from diffusion to kinetic control and at this point the reaction rate falls dramatically.

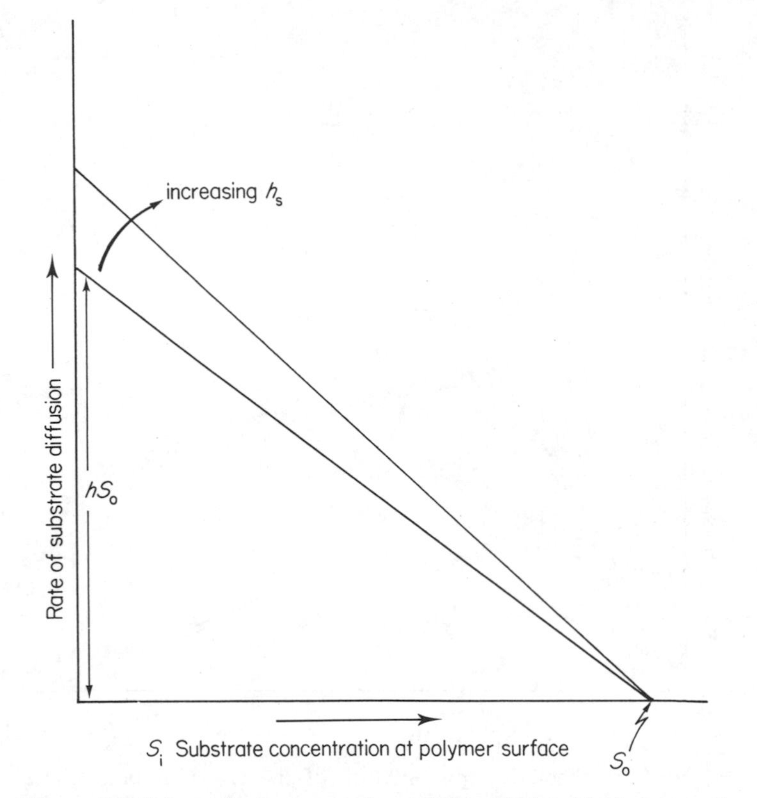

Figure 30. Schematic representation of the relationship between S_i and the rate of substrate diffusion to the surface of the polymer matrix, for a fixed value of S_0 and varying values of the substrate transport coefficient (h_s)

If S_0 is now reduced, because the reaction is kinetically controlled and the enzyme activity is low, the rate remains low until a point is reached where the reaction rate has increased sufficiently and S_0 reduced sufficiently so that substrate diffusion becomes the limiting factor. At this point the reaction rate rises rapidly whence further reductions in S_0 lower the reaction rate. The biological significance of hysteresis loops is discussed later (p. 120), but it is important to note here that at relatively high substrate concentrations the presence of diffusional restrictions will result in an increased reaction rate. Thus the diffusional limitation when combined with substrate inhibition, unlike other types of inhibition, will augment the rate of the reaction.

One question remains; where multiple steady states are permissible, when the immobilized enzyme is presented with a fixed rather than changing bulk phase substrate concentration, how does it 'know' which of the possible steady states to choose? To answer this question we must consider what happens when immobilized enzyme and substrate are mixed. In all that follows we will assume a fixed value of S_0 of 3 (dimensionless units see Figure 32). An enzyme

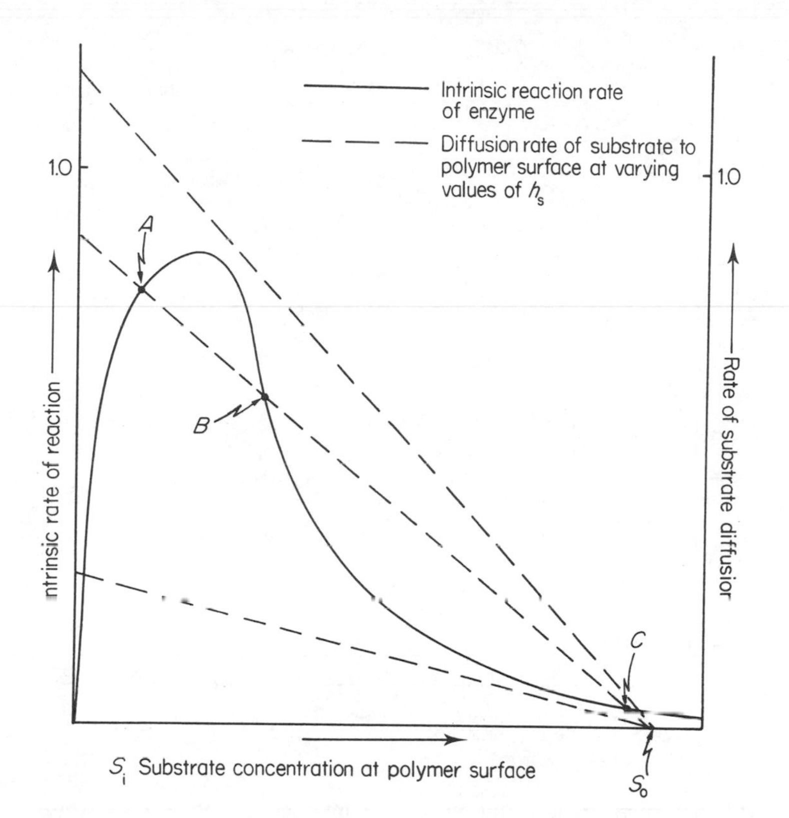

Figure 31. Schematic representation of the interaction between substrate diffusion limitation and substrate inhibition on an immobilized enzyme, for a fixed value of S_0 with varying h_s. Steady states occur when the rate of substrate diffusion to the polymer surface equals the intrinsic rate of the immobilized enzyme reaction

reaction may be initiated in one of two ways, either a solution of substrate and a solution/suspension of enzyme and all necessary activators may be mixed, or else substrate and enzyme may be allowed to equilibrate and the reaction initiated by the addition of an activator (or second substrate). In the former case, before the reaction can commence, the substrate must diffuse to the enzyme. Thus S_i starts at a value of zero, the rate of substrate transport is high, and the reaction rate low. As substrate diffuses to the polymer surface S_i increases, the transport rate decreases and the reaction rate increases until the two equate (intersection *A*, Figure 32). In the case where enzyme and substrate are equilibrated before the reaction is initiated S_i is equal to S_0, the transport rate is zero and the reaction rate immediately after initiation is the same as the free solution enzyme at a concentration of S_0. As the reaction proceeds S_i declines, the transport rate increases more rapidly than the enzyme reaction rate until the two become equal (intersection *C*, Figure 32). Thus the experimental method will determine the steady-state rate obtained. The steady-

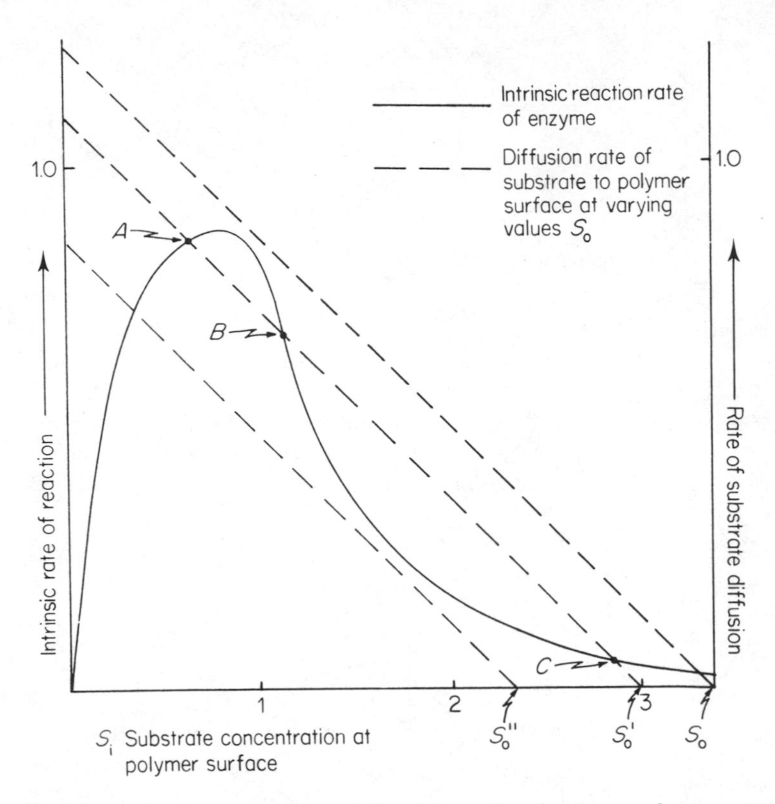

Figure 32. Schematic representation of the interaction between substrate diffusion limitation and substrate inhibition on an immobilized enzyme for fixed value of h_s with varying S_0

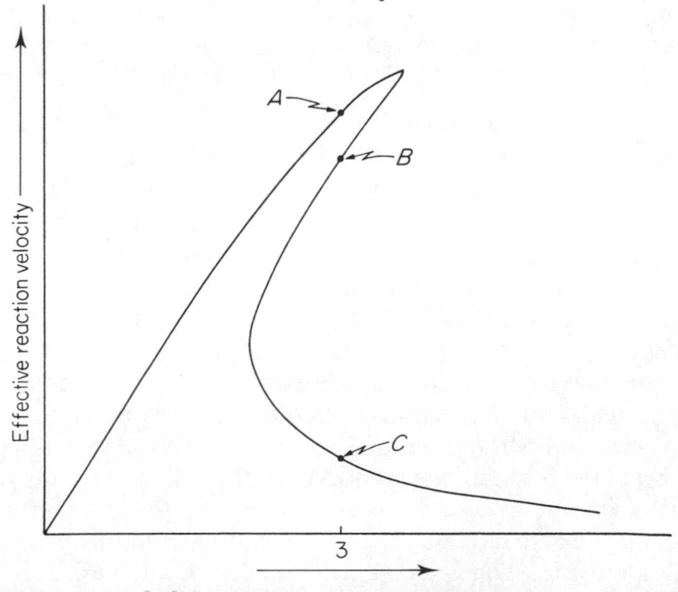

Figure 33. Schematic representation of the relationship between S_0 and the effective rate of reaction of an immobilized enzyme subject to substrate inhibition and diffusion limitation. Points *A*, *B*, and *C* are the steady-state rates indicated thus in Figure 32

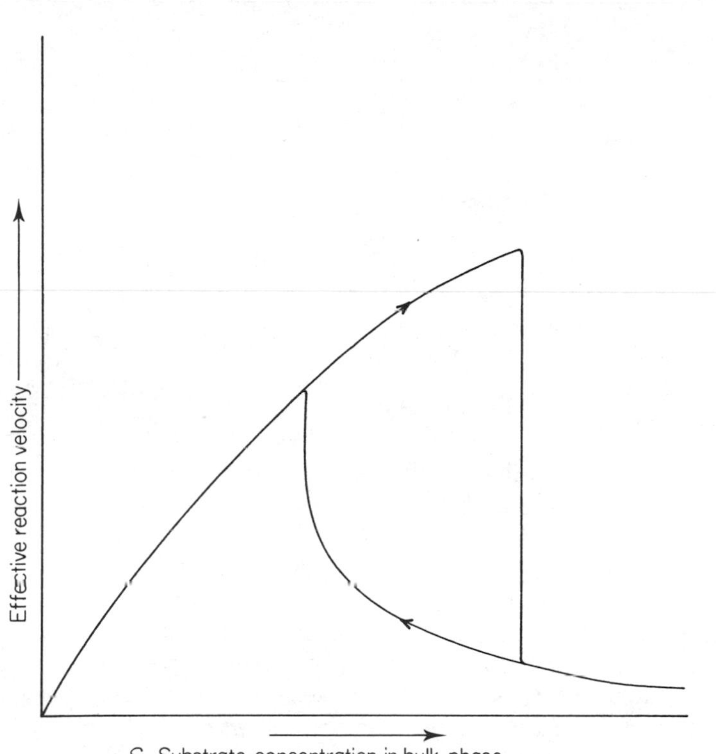

Figure 34. Hysteresis relationship between S_0 and the effective rate of an immobilized enzyme reaction subject to substrate inhibition and substrate diffusion limitation. Arrowheads denote the direction of change of S_0

state rate characterized by intersection *B* (Figure 32) will never be observed as it represents an unstable state. As a matter of interest a reinspection of Figure 31 will reveal that when $h_s = \infty$, i.e. no diffusional resistance is present, the immobilized enzyme will behave like the free solution enzyme.

Chapter 3

Application of Immobilized Enzymes

Ever since their initial development, the possibility of the use of immobilized enzymes as practical re-usable catalysts has been recognized and an enormous quantity of research has been carried out in this area. However, despite continual promises, the 'golden egg' has yet to be laid. The practical applications of immobilized enzymes may be divided into three main categories; analytical uses, therapeutic uses, and preparative uses.

Three criteria are pertinent when considering the use of an immobilized enzyme system. First, is the immobilized enzyme system cheaper than the existing system? Second, is the immobilized enzyme system more effective than the existing system? Third, is the process possible with any other system? The overriding consideration in the case of preparative uses is cost. In the case of therapeutic uses the most important feature of a potential immobilized enzyme system is that it shall be more effective than the existing treatment. In the case of analytical applications either a substantial cost or effectiveness advantage will be the deciding factor. One aspect of immobilized enzyme processes that makes them highly practical is their potential for automation. The importance of this and their other relative advantages will depend upon the economic climate at the time; when labour is cheap, who can afford automation? More detailed consideration of such economic factors is presented throughout this chapter, but especially in Section IV.

I. ANALYTICAL USES

The analytical applications of immoblized enzymes may be divided into two areas, the enzyme electrode and automated analysers. The distinction between enzyme electrodes and automated analysers may not at first be obvious, as many enzyme electrodes may be incorporated into automated systems. It will be useful therefore to define the two terms. Enzyme electrodes are probes capable of generating an electrical potential as a result of a reaction catalysed by an immobilized enzyme that is fixed on to or around the probe. The use of immobilized enzymes in automated analysis implies that the immobilized enzyme is used to replace soluble enzyme in an existing automatic analyser system.

1. Enzyme Electrodes

Many enzyme electrodes have been designed, but relatively few are in common use and fewer still in commercial production, as most have serious limitations in terms of cost, convenience, or life expectancy.

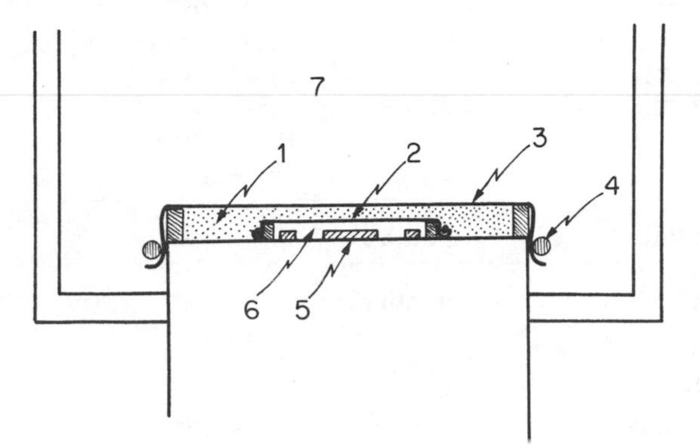

Figure 35. Diagram of a glucose-sensitive enzyme electrode. (1) Immobilized glucose oxidase. (2) Teflon membrane. (3) Cellulose acetate membrane. (4) 'o'-ring seal. (5) Platinum oxygen electrode. (6) Saturated KCl solution. (7) Sample solution

The first enzyme electrode to be prepared was a glucose-sensitive electrode (Figure 35) made by immobilizing glucose oxidase around an oxygen electrode. The glucose oxidase was immobilized in a polyacrylamide gel and held in place around the electrode by a piece of cellulose acetate. The principle of operation is extremely simple; the enzyme catalyses the removal of oxygen from solution at a rate dependent upon the concentration of glucose present. The electrode is inexpensive to make providing that an oxygen electrode is available, easy to use particularly with opaque or particulate solutions, highly specific, but requires frequent standardization. It is not, however, a viable proposition for commercial production and hence must be made in the laboratory in which it is to be used; this may be a serious limitation, particularly as many other routine methods are available (some involving commercially produced immobilized enzyme systems see p.59.

The general principle of enzyme electrodes is embodied in the glucose electrode. Whether or not a particular enzyme electrode is practicable will depend upon four factors, sensitivity, stability, response time, and cost. The glucose electrode of the type described can remain stable (i.e. usable) for over 4 months, has a sensitivity range of 10^{-1} to 10^{-5} mol l^{-1}, a response time of about 1 minute, and is cheap. It is, however, not in routine use because it takes time to set up and is difficult to automate. In an analytical laboratory such time is often not available as are alternative automated techniques. Many

Table 3 Enzyme electrodes

Type	Sensor Electrode	Stability	Range (mol l^{-1})	Response Time (Min)
Urea	Cation	4 month	10^{-2}—10^{-4}	1-2
	pH	3 week	10^{-3}—10^{-5}	5-10
	Gas (NH_3)	4 month	10^{-2}—10^{-5}	2-4
Glucose	Pt (O_2)	4 month	10^{-1}—10^{-5}	1
	Pt (H_2O_2)	14 month	10^{-2}—10^{-4}	1
	Gas (O_2)	3 week	10^{-2}—10^{-4}	2-5
	pH	1 week	10^{-1}—10^{-3}	1-10
Amino acid	Pt (O_2)	6 month	10^{-3}—10^{-5}	0.2
Alcohol	Pt (O_2)	4 month	1—100mg%	0.5
Penicillin	pH	2 week	10^{-2}—10^{-4}	1-2
Amygdalin	Cyanide	1 week	10^{-1}—10^{-5}	1-3

other electrodes have been constructed and a list is given in Table 3. It can be seen from this table that the principle of using a probe sensitive to some reactant or product of the reaction is widely applicable. There is one major drawback and that is the selectivity of the sensor probes, for example it is quite feasible to immobilize urease around an ammonium-ion-sensitive electrode, and produce a urea electrode. However ammonium-ion electrodes are also sensitive to potassium ions and unless the solutions of urea for assay have a known, constant potassium concentration the system will be impractical in use. When the electrode to be used is of the flat bottom type, then the mode of enzyme entrapment used in the first glucose electrode, polyacrylamide gel entrapment, is generally applicable. This method has the particular advantage of giving large quantities of enzyme activity. Many electrodes used are of the conventional 'dip' type and it is not easy to tie a thick suspension of polyacrylamide gel evenly around such an electrode with a piece of cellulose acetate film. An alternative technique for coating such electrodes is available. It first involves casting a thin layer of cellulose nitrate around the electrode. This is performed by rotating the electrode in a horizontal position and pouring over it a solution of cellulose nitrate in ether/ethanol. The washed electrode is then left to soak in a solution of enzyme and the adsorbed enzyme cross-linked in place with glutaraldehyde or some other bifunctional reagent.

2. Autoanalysis

There are two areas of autoanalysis in which immobilized enzymes may find general application, repeated autoanalysis of small samples (e.g. blood samples) and continuous stream monitoring of large volumes.

The conventional autoanalyser in use in laboratories for the repeated enzymic analysis of small biological samples suffers from one major disadvantage, an aliquot of enzyme must be added to each sample and is subsequently lost. The alternative is to immobilize the enzyme and fix it in the sample stream. Plugging the sample stream with a column of immobilized enzyme would create too much resistance to the flow to be a useful approach. However, the enzyme may conveniently be attached to the inner wall of a narrow bore nylon tube, through which the sample stream is made to pass. Obviously the precise dimensions of the tube in relation to flow rate have to be carefully calculated to afford the optimum reaction system. Such a device may be prefabricated and a coil tube of immobilized hexokinase and glucose-6-phosphate dehydrogenase for the autoanalysis of glucose concentrations has recently been marketed. Another virtue of such an immobilized enzyme system is that more than one determination could be performed on each sample. Figure 36 illustrates the principle involved. The sample stream flows first through one enzyme coil, the reaction is monitored and as there is no contaminant enzyme, the sample stream can be made to flow through another enzyme coil and monitoring system.

Where there is a large volume of material to be analysed continuously another approach may be adopted. The principle involved is the measurement of heat generated by an enzyme reaction and was first reported by Mosbach and Dannielson. If a thermistor is held in the middle of a column of immobilized enzyme as a substrate solution is passed through the column, the reaction catalysed around the thermistor raises the temperature. Given that the flow rate, initial substrate solution temperature, and other physical parameters remain constant, the rise in temperature created by the reaction will be a measure of the substrate concentration (see Figure 37). If the solvent flow always contains substrate then a continuous curve will be produced. If substrate is introduced intermittently into the solvent flow then intermittent rises in temperature will be recorded. Such an analytical system might be set up to draw off part of the output stream of a continuous industrial process, in order to monitor the concentration of the product (Figure 38). Alternatively such a system, containing an appropriate (mixture of) enzyme(s), might be used as a check for contaminants in a product stream.

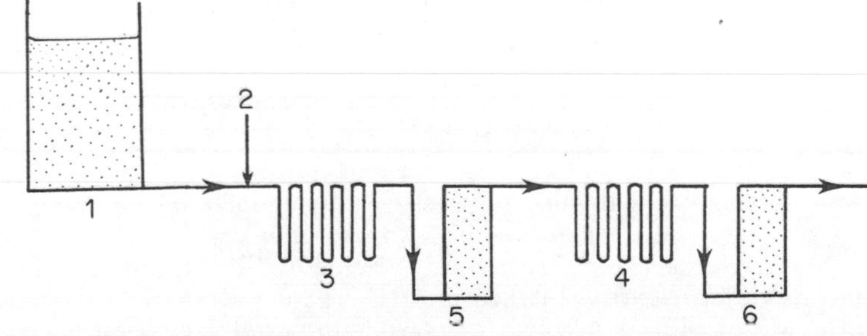

Figure 36. Illustration of the principle of the use of immobilized enzyme coils in automated multiple sample analysis. (1) Buffer reservoir. (2) Sample injection. (3) First enzyme coil. (4) Second enzyme coil. (5) and (6) Cuvettes

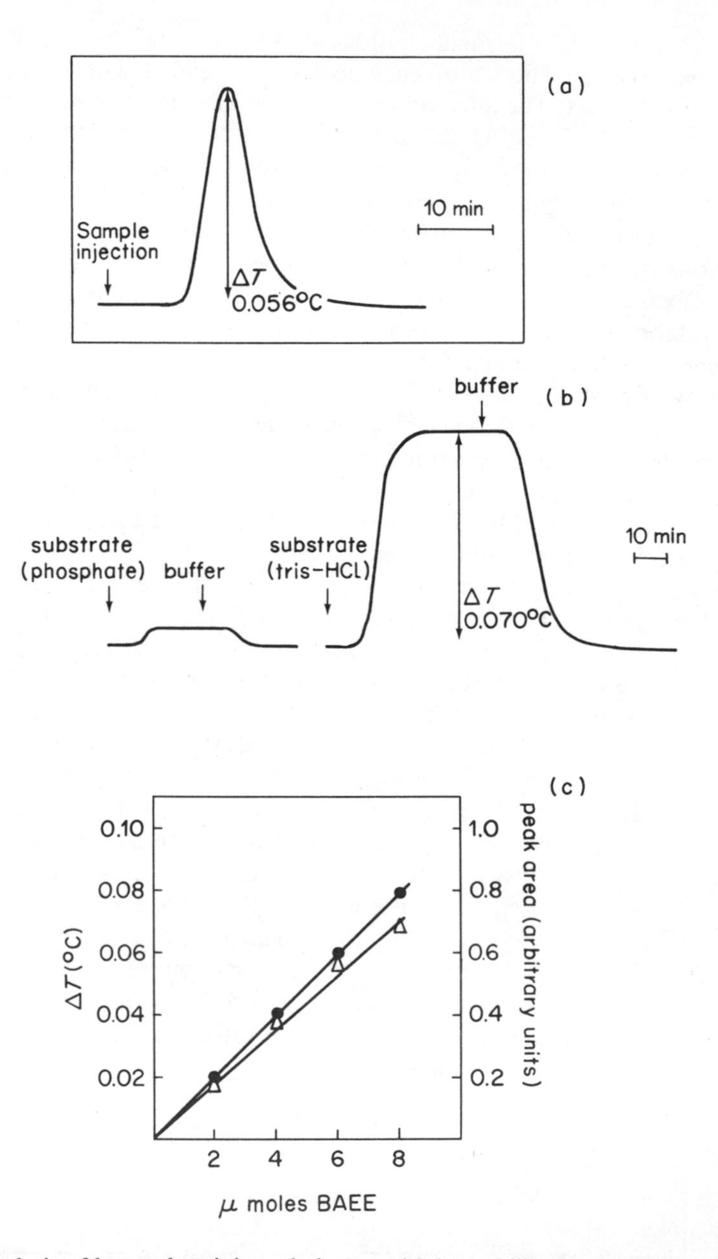

Figure 37. Analysis of benzoyl arginine ethyl ester with immobilized trypsin thermistor column. (a) Curve obtained after injection of a pulse of BAEE into the continuous buffer flow. (b) Curve obtained by continuously passing BAEE through the column in phosphate or Tris buffer. At arrow marked 'buffer' column continuously flushed with pure buffer. (c) Measured peak heights ($\triangle T$) ($\triangle$) and integrated peak areas (●) as a function of the amount of BAEE pulse injected into the immobilized trypsin thermistor column. Reproduced with permission from Mosbach and Dannielson, *Biochim. Biophys. Acta*, **364**, 140 (1974)

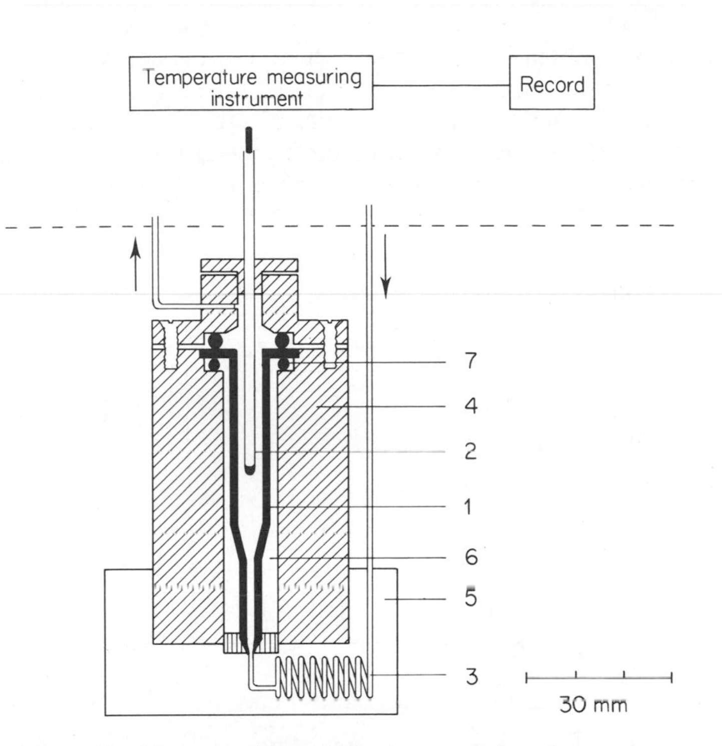

Figure 38. Diagram of immobilized enzyme thermistor column. (1) Microcolumn of immobilized enzyme. (2) Thermistor. (3) Heat exchanger. (4) Perspex cylinder. (5) Water jacket. (6) Air space. (7) 'O'-ring seal. The entire unit is placed in a water bath to the level of the dotted line. The arrows indicate the direction of the substrate flow. Reproduced with permission from Mosbach *et al.*, *Biochim. Biophys. Acta*, **403**, 256 (1975)

Another general advantage of this type of system is that it does not require the solvent stream to be optically clear. Thus it may be adapted for monitoring substances in blood (or other biological fluids). A system for monitoring glucose in serum is in use in Sweden. Solvent flows continuously through the column and small samples of serum are injected into the top of the column. Transient temperature changes within the column are recorded and compared with those produced from standard samples. This process once set up and stable (stabilization may take upwards of 2 hours) is thus simple and rapid to use.

Often the substance to be analysed is not susceptible to enzymic catalysis, but may well be an enzyme inhibitor. A variation on the above approach may then be used. Let us take the example of a waste solvent flow which might contain mercury salts as a contaminant. A column is constructed containing immobilized acetyl cholinesterase (subject to inhibition by mercury). Waste solvent flows continuously through the column. Acetyl choline (the enzyme's substrate) is intermittently injected into the top of the column. If no mercury is present then the temperature rise recorded for each substrate addition is identical and characteristic of the quantity of substrate added. If mercury is introduced into the solvent flow then the temperature rise is reduced by an amount proportional to the degree of enzyme inhibition and hence mercury present (see Figure 39).

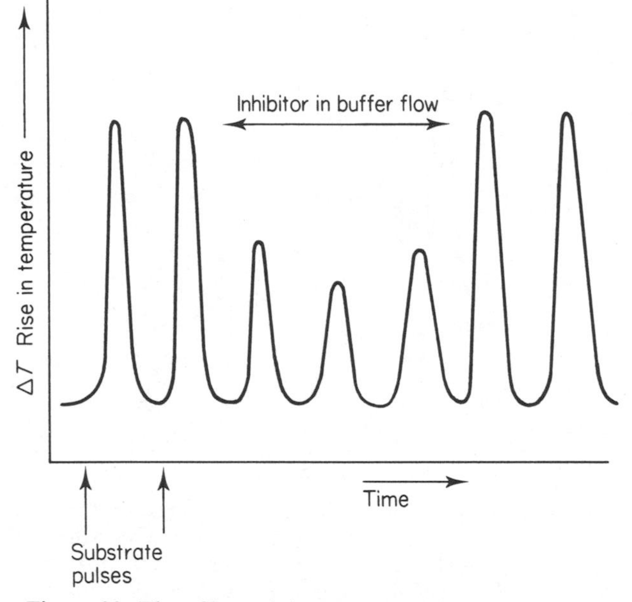

Figure 39. The effect of the presence of inhibitor in the buffer stream on temperature rise in an immobilized enzyme thermistor column. Substrate pulses are injected at regular intervals

The technique has two major advantages, it may be used in an automatic quality control role and, the analysis is immediate. As a generally applied technique, however, it does suffer from several major drawbacks. The temperature changes in the column are very small and thus precise environmental temperature control is necessary. This is usually achieved by a sophisticated water jacket system which controls the temperature of the solvent flow and column. Flow rates must be accurately maintained and the column requires frequent standardization. All this naturally costs money. The cost of the hardware of a column used in the Swedish glucose analyser is of the same order as a medium priced analytical high speed centrifuge. Two other problems also exist. If the substrate is being introduced intermittently into the column, then it must be dissolved in the same solvent system as the bulk solvent flow to avoid changes in temperature due to solvation effects. It is also possible to upset the system if the thermal capacity of the solvent flow is altered for any reason. Nevertheless, despite such complications this will undoubtedly prove to be a useful analytical technique.

II THERAPEUTIC USES

Enzymes may be employed as therapeutic agents, either to replace enzymes that are missing because of genetic or other malfunctions in the individual or as agents for the specific degradation of unwanted metabolites in the diseased individual.

1. Enzyme Replacement

There are many diseases that are characterized by the absence of a particular enzyme either because of genetic malfunction or because of tissue malfunction. Examples of genetically derived enzyme deficiency include the broad classification of lysosomal storage diseases, where a lysosomal enzyme is absent and its substrate will accumulate within the cell, often with disastrous results and also diseases such as Favism where an enzyme involved in a metabolic pathway, in this case glucose-6-phosphate dehydrogenase of the red blood cell, is absent.

Tissue malfunction, particularly of the liver and kidney, may result in the accumulation of toxic waste materials, for example urea, within the body.

Such diseases can be controlled, at least in theory, by injecting soluble enzyme from an extraneous source (e.g. microbes). This approach may cause more problems than it solves; a foreign enzyme protein will almost certainly cause an allergic response that in itself may be fatal; the soluble enzyme may be unstable and will be rapidly cleared from the body by the natural immune response system. Immobilization may overcome both of these problems, by preventing interaction between the enzyme and the body's immune response system and stabilizing and protecting the enzyme. Potentially the most useful form of immobilization is encapsulation within a non-antigenic polymer material, e.g. nylon or collodion. The exact procedure chosen will depend, however, on the mode of administration of the immobilized enzyme, whether it will be by direct injection when a biodegradable polymer is indicated, or by extracorporeal shunt when vinyl polymers may be used.

Whereas the majority of enzyme replacements will involve simple hydrolytic or oxidase reactions (e.g. urease or catalase), certain enzyme deficiency diseases involve dehydrogenases or kinases. In these cases NAD/NADH or ADP/ATP regenerating systems must be included with the replacement enzyme, because coenzymes are not generally available in the extracellular compartment. For example, the inherited galactokinase deficiency disease may be treated (experimentally) by injecting galactokinase coencapsulated with an ATP regenerating system of ATP/ADP and pyruvate kinase.

2. Routes of Administration

The manner of administration of the enzyme will depend, in part at least, on the nature of the disease, in particular where in the body the substrate for the enzyme is to be found, and how rapidly the toxin reaches harmful levels in the untreated diseased state. Where the substrate is present in the extracellular fluid, for instance urea in liver/kidney malfunction, the immobilized enzyme may either be applied by injection or included in an extracorporeal shunt, the choice depending upon the rate of accumulation of the substrate. Obviously the use of an extracorporeal shunt will only be appropriate where the accumulation of substrate is sufficiently slow to allow it to be cleared on a weekly or monthly, rather than daily, basis. Where continual removal of the substrate is indicated then injection must be used. If the substrate is contained entirely intracellularly, as is usual in lysosomal storage diseases, then not only must the immobilized enzyme be administered by injection, but must preferably be targeted to be absorbed specifically by the diseased cell. We shall discuss in turn administration by injection, injection with targeting, and extracorporeal shunts.

Injection of non-allergic immobilized enzymes intravenously usually results in their rapid accumulation in the liver and spleen where they are quickly degraded. An additional problem exists in that particle size must be accurately controlled to avoid blockage of capillaries. However, direct intravenous injection may still be desirable, as seen in the suggestion of the use of artificial red blood cells produced by encapsulating haemoglobin within a cellulose-based polymer or polylactic acid capsules. Subcutaneous injection is not ideal because of the slow rate of equilibration between this compartment and the remaining extracellular fluid and the obvious limit to the quantity of enzyme that can be administered. Thus the intraperitoneal cavity is the favourite site of administration. Particles of a diameter greater than 20 μm are retained more or less indefinitely within the peritoneal cavity and experiments have demonstrated that effective enzyme activity may be retained for up to 8 months. However, certain solutes will not cross into the cavity and thus this route cannot be used in all cases.

When the biochemical lesion is a totally intracellular event, for instance in lysosomal storage diseases where the absence of an enzyme leads to an accumulation of substrate entirely within the diseased cell, then replacement of the enzyme by inclusion in some compartment of the extracellular fluid will be ineffective. Another approach must be adopted which will lead to the replacement enzyme being adsorbed specifically by the diseased cell. The only practical approach is to introduce into the bloodstream the immobilized enzyme which has been programmed to attach itself specifically to the target cell. The most hopeful form of immobilization in this case is liposomal entrapment of the enzyme. Unmodified liposomal entrapped enzymes tend, however, to be cleared from the circulation by the liver and spleen and as yet, despite much research, no method has been found for modifying the liposomes so that they will be specifically absorbed by other defined tissues. In theory, at least, it should be possible to programme the liposomes by making use of the unique cell surface markers of different tissues, perhaps by including specific antibodies to these markers on the surface of the liposomes.

Extracorporeal shunts, where the blood is brought into contact with the appropriate immobilized enzyme outside the body, may be used to eliminate unwanted substrates from the bloodstream in cases where some tolerance exists within the individual. Artificial kidney machines are an example of a non-enzymic extracorporeal shunt, designed to rid the blood of, among other things, urea by a process of ultrafiltration. Urea could be broken down enzymically in an extracorporeal shunt consisting of a column of immobilized urease. In practice however, this would pose an alternative problem, how to dispose of the toxic ammonium carbonate produced from urea hydrolysis! Nevertheless the general application of the technique seems promising for a variety of reasons, the most obvious of which is that foreign substances are not introduced directly into the body. Other advantages also accrue from the use of an extracorporeal shunt system. The method or polymer material for immobilization is less important, for instance it would be possible to immobilize the enzyme by microencapsulation in nylon or by attaching it to the inner wall of several metres of coiled nylon tubing. This latter approach has the added advantage of excellent, unrestricted flow characteristics. In general, extracorporeal shunts can only be used where the product of the reaction is non-toxic, rapidly detoxified or excreted by the body or where some detoxification step is included in the shunt.

3. Enzyme Therapy

Enzyme therapy differs from enzyme replacement in that the enzyme to be added to the body is not one that is normally found within the body (or is present but not pathologically diminished in quantity). The rationale behind the addition of such an enzyme is to alter the normal environmental conditions in the body in order to control a diseased state. The best example of this is the use of asparaginase in the treatment of certain leukaemias. Normal cells have the ability to synthesize asparagine, but certain leukaemic cells do not and without an extraneous supply from the blood cannot grow. The addition of immobilized asparaginase to the blood or contiguous extracellular compartment reduces the blood asparagine concentration to a minimal level and the leukaemic cells die from asparagine starvation. The asparaginase must be immobilized in a biodegradable form, for example by inclusion in a liposome or polylactic acid capsule. Administration can be either intravenous or intraperitoneal.

III INDUSTRIAL AND PREPARATIVE USES

When considering the application of immobilized enzymes to industrial processes it is prudent to remember the maxim 'all that is possible is not necessarily worth while'. Any industrial process is ultimately concerned with producing a product of the highest possible quality for the lowest possible price. Although many potential industrial applications of immobilized enzymes exist, very few operate commercially. As economic pressures and realities change, new applications are liable to be instituted. The rest of this chapter will describe present-day applications of immobilized enzymes, the economic and practical problems involved in setting up an immobilized enzyme process, and finally some speculation on possible future developments.

Why use immobilized enzymes in industrial processes? To answer this question we must look first at the reasons for using enzymes as industrial catalysts. The end to be achieved is the chemical transformation of a substance to provide a useful product. Three approaches are available: (1) purely chemical transformation, (2) fermentations using whole microbial cells, and (3) reactions catalysed by isolated enzymes or crude cellular extracts of enzymes. The theoretical ability of the organic chemist to perform almost any chemical transformation is virtually infinite, however, in practice many chemical transformations are either too costly or technically difficult to render pure chemistry of any commercial value. For example, the chemical synthesis of antibiotics is impracticable, as would be the chemical removal of lactose from milk. Cellular fermentations have the advantage, unlike many chemical procedures, of ensuring a certain specificity of reaction, generally requiring normal temperatures and pressures and providing high yields of product. They do, however, have several drawbacks; undesirable products may be formed because alternative metabolic pathways may be present in the cell: rigorous aseptic conditions must be used. There are major advantages of using isolated enzymes or crude cellular extracts, rather than whole cells: an enzyme reaction is usually specific both in terms of substrate and product; microbial contamination of the product is less of a problem; no special medium to feed the cells is required; there is no cell membrane to prevent interaction of enzyme and substrate. A number of major problems are encountered in using enzymes as industrial catalysts: the enzyme may be unstable; it is not practically viable to recover a soluble enzyme; batch processing must usually be employed and, as a result, processes based on soluble enzymes are difficult to automate. Some or all of these problems may be overcome by immobilizing the enzyme: it may be rendered more stable; it can be easily recovered and re-used; continuous production processes are possible which are amenable to automation. These and other factors are discussed later in this chapter. Thus employing immobilized enzymes is in most cases a logical extension of the use of soluble enzymes. It will be appropriate first to describe briefly the hardware of enzyme technology, the enzyme reactor, before investigating details of individual processes.

1. Enzyme Reactors

The purpose of an enzyme reactor is to allow enzyme and substrate to come into contact for a sufficient period of time for the reaction to take place and then to be able to separate easily the product and enzyme. Several configurations of reactor are feasible and we shall discuss each in turn. Schematic illustrations of each will be found in Figure 40.

a. Batch Reactors

Batch reactors are essentially large stirred tanks into which are placed enzyme and substrate. The reaction is allowed to proceed to completion, the reactor drained, and the product separated from the enzyme. If a soluble enzyme has been used it is usually removed by denaturation (e.g. heat treatment). This procedure is economically feasible so long as the enzyme is cheap. Potentially at least, expensive enzymes may be used in a batch reactor if they are first immobilized and the immobilized enzyme subsequently recovered by centrifugation or filtration. In practice however, the immobilized enzyme may be destroyed by the recovery process and although batch reactors are often used in conjunction with (cheap) soluble enzymes, their use with immobilized enzymes has limited potential. The one major advantage of the batch reactor is that it is cheap in comparison to other types of reactor.

b. Continuous Flow Reactors

The principle of a continuous flow reactor is the continuous addition of substrate to and removal of product from the enzyme reactor. There are a number of ways in which continuous reactors may be designed and these are outlined below.

The *continuous flow stirred tank* reactor consists of a stirred tank with a separate substrate inlet and reactor mix (i.e. product) outlet. By judicious choice of tank size, enzyme activity, and rate of substrate addition, the percentage conversion of substrate to product may be adjusted. For example low flow rate, large reactor size (and hence long residence times), and high enzyme activity will give high product yields. These considerations apply equally to other reactor configurations.

The immobilized enzyme may be retained within the reactor by filtration of the product stream, incorporating a subsequent settling stage, immobilizing the enzyme on to a magnetically active particle and retaining it within a magnetic field (which may also be used to stir the particles) or by immobilizing the enzyme to the paddles of the agitator shaft.

Continuous flow stirred tank reactors may be combined with an ultrafiltration process that will allow the use of soluble immobilized enzymes within the reactor. This will be advantageous when the substrate is itself insoluble or colloidal.

Immobilized enzyme particles lend themselves to being packed in columns. Substrate may then be passed through the column over the enzyme and product obtained from the outlet. The idealized *packed bed reactor* is one in which the fluid velocity profile over a given transverse cross-section of the column is perfectly flat; such a system is known as a *plug flow reactor*. In practice the substrate stream may enter from either the top or bottom of the column and the column may be constructed as a tall column or a flat bed. As an alternative to using columns packed with immobilized enzyme particles, the enzyme may be immobilized to a membrane or sheet of material (e.g. filter paper) and confined in a filter press arrangement. Recently reports have been made of spiral membrane packed bed reactors, in which the enzyme is immobilized to a sheet of porous membranous material which is rolled into a spiral and inserted into a column.

Another variant of the packed bed reactor is the *hollow fibre reactor*. Hollow fibres, the walls of which are permeable to substrate and product but impermeable to enzyme molecules, are packed into a column. Enzyme may be retained inside the fibres and the substrate stream passed round the outside of the fibres. Substrate will diffuse through the fibre wall, react with the enzyme, and the product so formed will diffuse back through the fibre wall into the fluid stream. Alternatively, substrate may be passed through the centre of the fibre which is bathed by a solution of enzyme. *Fluidized bed reactors* are a hybrid of continuous flow stirred tank and packed bed reactors. The immobilized enzyme is loosely packed into a column and the substrate stream passed into the bottom of the column at a sufficiently high flow rate to lift and mix the immobilized enzyme particles within the column. Obviously the immobilized enzyme particles must have a higher density than the reaction

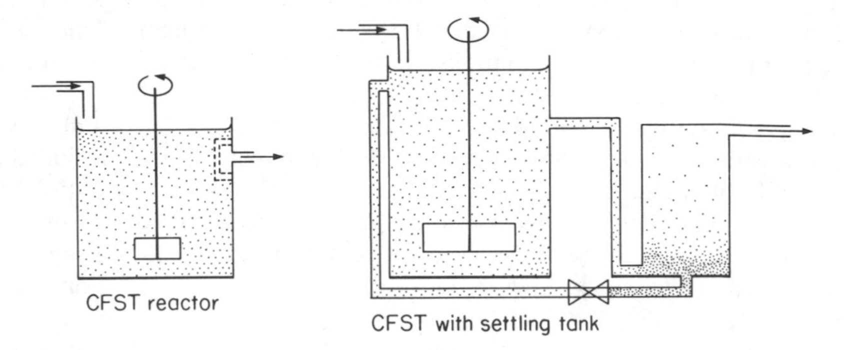

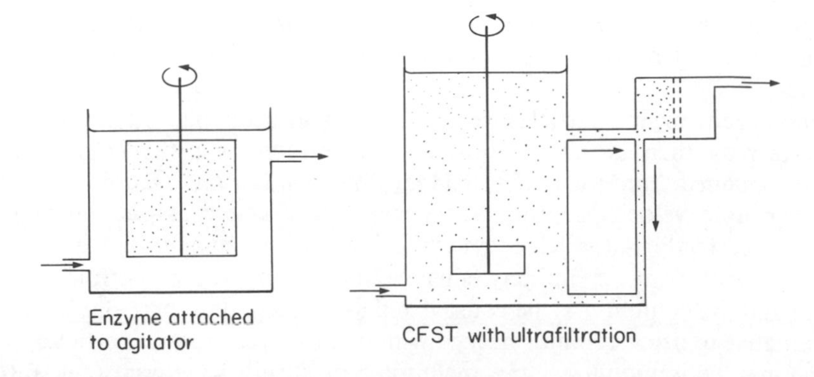

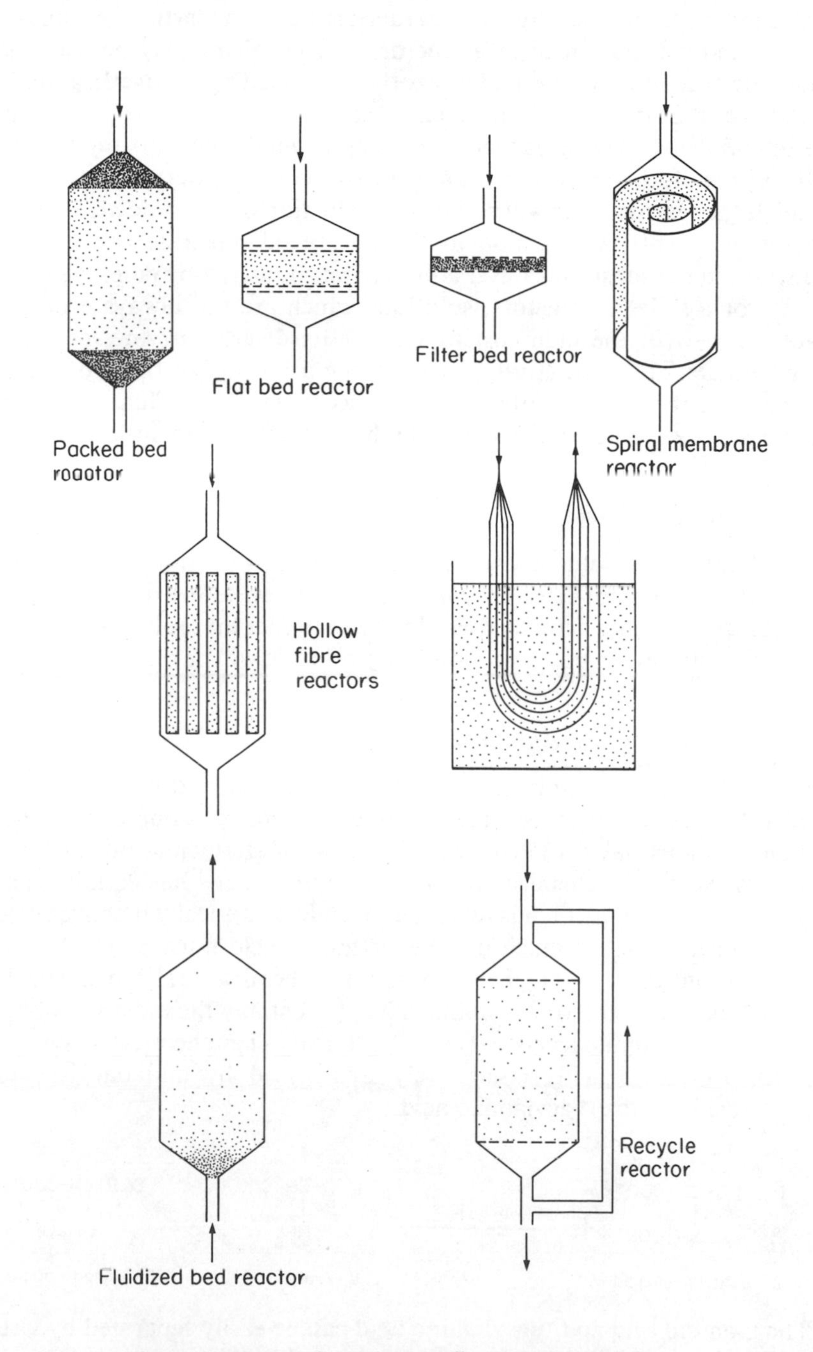

Figure 40. Types of continuous flow enzyme reactor

medium and the flow rate must not be so high that the enzyme particles are carried out of the top of the column.

Depending upon the physical characteristics of a particular immobilized enzyme and substrate used in a reactor, two problems may be encountered. First, the reaction may be highly exothermic, leading to overheating of the fluid stream. Second, large unstirred layers of solvent may be present around the immobilized enzyme particles resulting in inefficient mixing of substrate and enzyme. Both these problems may be overcome by increasing the flow rate through the reactor. This will result in a reduction of the residence time of the reactor and hence a reduction in the degree of substrate conversion. The reduction in the substrate conversion may be countered by employing a larger reactor or a series of reactors, solutions which are not always economically viable in view of the high cost of reactor hardware, enzyme, and polymer support material. Alternatively, by recycling a portion of the product stream through the reactor, the effective residence time of the fluid stream in the reactor may be increased, despite the high flow rate, at very little extra cost.

2. Industrial Processes

Despite the large volume of research work carried out on the potential applications of immobilized enzymes, relatively few industrial processes using immobilized enzymes actually exist. In the examples that follow only industrial processes currently in use or economically possible will be discussed.

a. The Resolution of DL-*Amino acids*

The growing use of L-amino acids in both medicines and human and animal foodstuffs has led to large-scale increase in demand. Amino acids may be produced industrially by either a process of fermentation or chemical synthesis. While the chemical synthesis of amino acids has certain economic advantages, the product tends to be the racemized, optically inactive DL-amino acid. In order to be of much use the DL-amino acid must be resolved to the L form. From the variety of methods, both chemical and biochemical, that exist for the resolution of DL-amino acids, potentially the most valuable is the use of mould aminoacylase (E.C. 3.5.1.14). The chemically synthesized acylated DL-amino acid is selectively hydrolysed by aminoacylase to the L-amino acid and the D-acylamino acid.

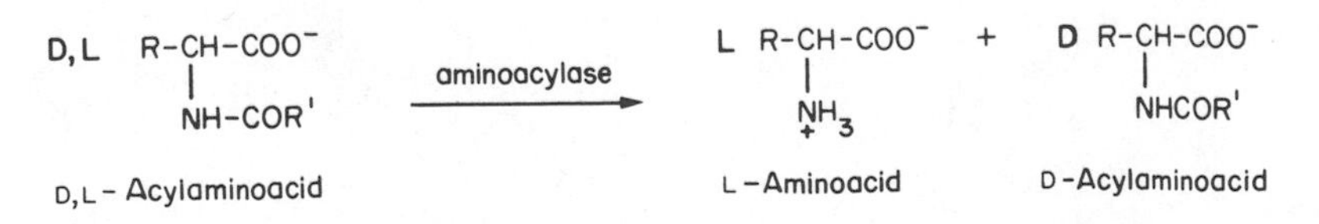

The L-amino acid and D-acylamino acid can be easily separated by virtue of their differing solubilities. The D-acylamino acid is then racemized to the DL form and used again.

Industrial processes for L-amino acid production based on the batch use of soluble aminoacylase were in operation as long ago as 1954, but they suffered from a number of drawbacks common to many batch-soluble enzyme processes, that is high labour costs, complicated separation of product and enzyme leading to increased cost and decreased yield, and non-reusability of enzyme. During the mid-1960s the Tanabe Seiyaku Co. of Japan were investigating ways of overcoming these problems by the use of immobilized aminoacylases. The industrial production of L-methionine by aminoacylase immobilized on to DEAE–Sephadex in a packed bed reactor, the first full-scale industrial use of an immobilized enzyme, was eventually set up by Tanabe Seiyaku Co. in 1969. Its great virtue is its relative simplicity and ease of control. Figure 41 is a flow diagram illustrating the principles of the production process. The half-life of the enzyme reactor was approximately 65 days at a temperature of 50 °C. The disadvantage of immobilizing the amino-acylase by physical adsorption on to DEAE-Sephadex were more than offset by the advantage of being able to regenerate the enzyme reactor by the addition of fresh enzyme, without having to discard the expensive DEAE–Sephadex. The productivity of a 1,000 litre column was between 200 and 700 kg h^{-1} depending upon the amino acid used. The process has subsequently been copied using a variety of support materials, both organic and inorganic.

b. High Fructose Syrups

The recent controversy over the safety of artificial sweeteners leading to the complete ban on cyclamates and the reduction in the use of saccharin in the food industry, has placed more emphasis on the traditional sweetener sucrose. While sucrose is sweet, fructose is approximately 1½ times sweeter and consequently a large quantity of invert syrups (i.e. hydrolysed sucrose) are produced commercially. Invert syrups are of course 50% glucose and 50% fructose and for a long time the food industry has been aware of the potential for using glucose isomerase to produce high fructose syrups from glucose. The Japanese were again first in the field, employing soluble glucose isomerase to produce high fructose syrups in 1966. The product of this reversible reaction is about 42% fructose, 50% glucose, and 8% other sugars. For a number of reasons, the economic being among the more important in this particular case, it was not until 1974 that the first commercial production of high fructose syrups using glucose isomerase immobilized on to a cellulose ion-exchange polymer, in a flat packed bed reactor, was initiated by Clinton Corn Products of Iowa. In this case the glucose syrup was obtained from corn starch, but in practice may be obtained from any form of cheaply available starch (e.g. potatoes). The potential market for high fructose syrups is large and has been estimated to be in the region of 2×10^6 tonnes per year.

c. Conversion of Starch to Glucose

In terms of sheer quantity the present industrial production of glucose syrups from starch is one of the major areas of biocatalysis. Some 10×10^6 tonnes of

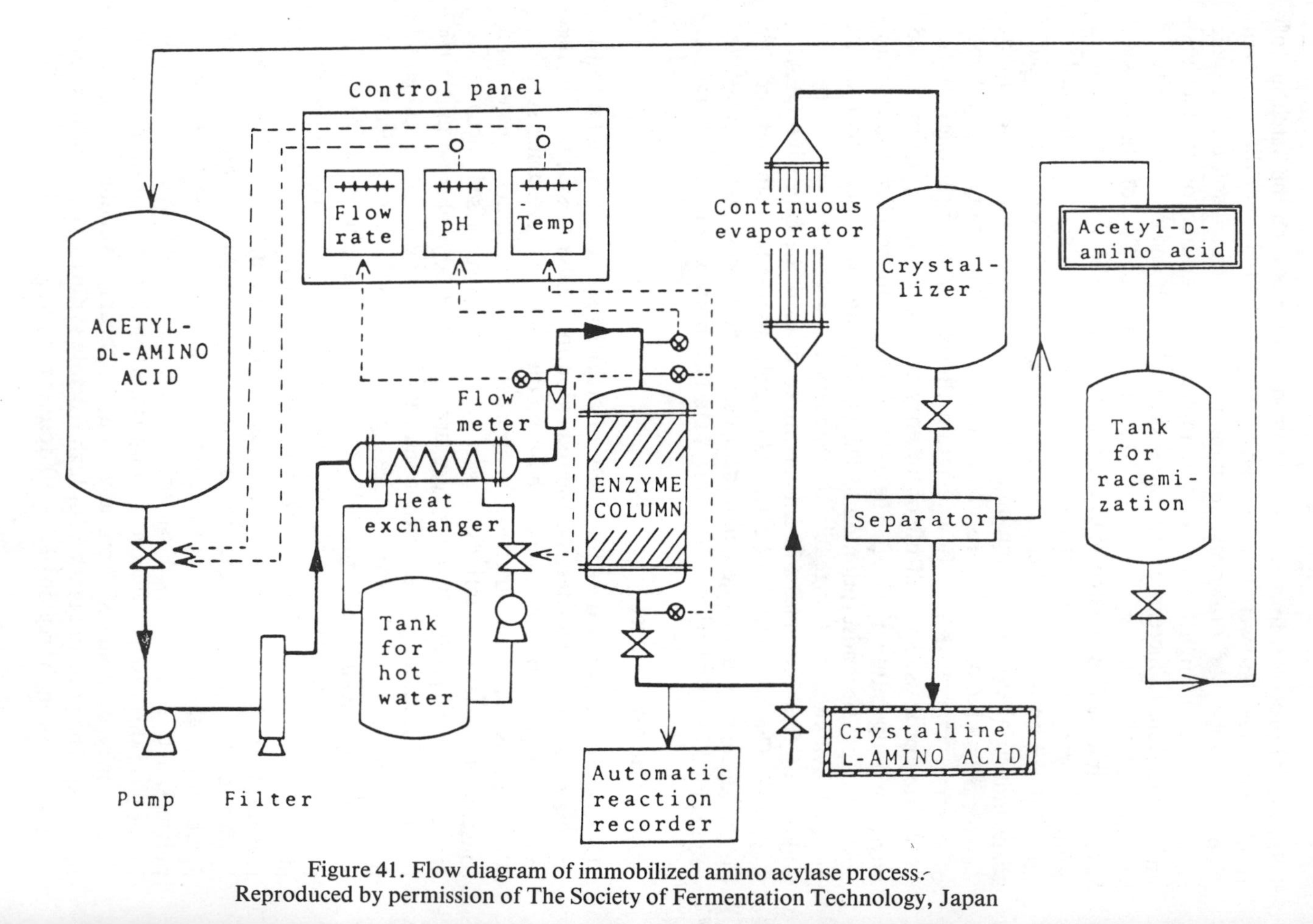

Figure 41. Flow diagram of immobilized amino acylase process. Reproduced by permission of The Society of Fermentation Technology, Japan

corn and glucose syrups are produced annually by batch conversion with soluble amyloglucosidase and, with the increasing requirement for glucose syrups as a substrate for the production of high fructose syrups, this demand must surely increase. Numerous attempts have been made over the past decade to develop an economic process based on immobilized amyloglucosidase using a continuous stirred tank reactor or a packed bed reactor, but as yet it would appear that none has been commercially adopted. One problem in this respect is that the present industry has the ability to increase its productivity without the need for new investment and it is likely that, until production by existing methods reaches the maximum existing industrial capacity, no new process will be implemented.

d. Treatment of Milk

The potential for the use of immobilized enzymes in the treatment of milk falls at present into three major areas, the production of cheese, enzymic stabilization, and the removal of lactose from milk products.

Cheese production is one of the most time honoured pieces of enzyme technology. The first stage in the production of cheese is the coagulation of milk, traditionally performed by adding soluble rennin isolated from the calf intestinal tract. The increasing cost of calf rennin has led to the introduction of the use of bacterial rennin. However, it would appear that the source of rennin used can affect the flavour of the cheese produced and not all bacterial rennins are suitable. A major cost element in cheese production is therefore the cost of copious supplies of the correct rennin and the capital costs of the large holding tanks in which coagulation takes place. Clearly a process based on the use of immobilized rennin or other suitable proteases would be of enormous economic value. One very obvious problem with such a process is the tendency for coagulation to occur within the immobilized enzyme reactor. It would appear that coagulation is temperature dependent and that skimmed milk may be passed through a packed bed reactor at low temperature and coagulation subsequently initiated by incubation for a few seconds at 37 °C.

The liquid waste from cheese production, the whey, contains protein, various salts, and large quantities of lactose. Not only are the hydrolysates of lactose, glucose and galactose, used in the dairy industry but they would provide a useful substitute for corn syrups for the production of sweeteners. To date, the enzymic hydrolysis of lactose has not been an economically viable proposition because of the prohibitive cost of lactase. Chemical hydrolysis is equally uneconomic because of the cost of purifying the products. Thus cheese whey is at present largely a waste product. Attempts have been made to process cheese whey using immobilized lactase but at present only a few small industrial plants exist. Even if a cheaper source of lactase could be found practical problems abound in the process, for example the prevention of bacterial contamination of the column and the clogging and rupture of the ultrafiltration membranes required to remove the protein in the product.

These latter problems have been a matter of quite considerable investigation, because they can affect many other potential immobilized enzyme processes, and are discussed separately.

The great attraction in developing a process for the continuous hydrolysis of lactose in whey remains the abundance and cheapness of the raw material and the potential value of the product. In fact acid whey, a by-product of cottage cheese, is usually disposed of with some difficulty and cost to the dairy industry.

There is, however, another good reason for the development of an immobilized lactase system and that lies in the inability of many individuals to metabolize lactose, because they lack the lactase enzyme. The clinical result of milk consumption by such people is severe gastrointestinal upset, diarrhœa, flatulence, and abdominal pain. Whether the deficiency of lactase is a genetic abnormality is not clear, but it is particularly widespread among the inhabitants of the Third World. This is perhaps even more unfortunate, since it is precisely such populations that could benefit most from the consumption of milk, a cheap, virtually complete food. The objective then is to be able to remove, cheaply, the lactose from milk without affecting its nutritional or physical characteristics. The most obvious method would be a one-step process based upon an immobilized lactase preparation. Several problems relating to such a procedure must be immediately apparent.

Milk is a colloidal fluid, therefore if a packed bed reactor containing immobilized lactase were to be used there would exist the possibility of fouling the column. In order to preserve the quality of the milk and prevent microbial contamination, sterilized skimmed milk must be used at a low operating temperature. This low temperature requirement will necessitate the use of either high loadings of highly active enzyme or long reactor residence times. Enzyme or polymer support leakage from the reactor must be at an absolute minimum in order to prevent expensive subsequent processing of the milk product. Happily the solution to these problems has been described and an industrial plant in Italy, with a minimum capacity of 8,000 litres a day, has been in operation since 1975. The nature of the reactor and process is interesting because it is an excellent example of how the varied problems of setting up an industrial immobilized enzyme process may be overcome. The high cost of the enzyme lactase (β-galactosidase from yeast) means that high operational stability must be achieved in order for the process to be economic. Given high operational stability, packed bed reactors have the advantage of ease of automation and control and thus, in a new industrial venture, will result in lower labour costs. They also require less enzyme for the same level of product yield (see p.86). In order to overcome the potential clogging of a packed bed reactor by the milk, relatively high column flow rates must be used and the immobilized enzyme must be packed at the minimum possible density. Thus either a highly active enzyme preparation must be used or a recycle reactor configuration adopted to allow the necessary 75% lactose hydrolysis to occur. In view of the cost of the enzyme the latter solution was adopted and a packed bed recycle reactor was designed (see Figure 40) to utilize an unusual but potentially highly useful form of immobilized enzyme, droplets of lactase trapped inside cellulose triacetate fibres (see p.8). The fibres are arranged in skeins and packed loosely in longitudinal bundles in a reactor column. Such an arrangement provides both low resistance to fluid flow and very high surface areas of immobilized enzyme which, combined with low temperature operation, minimizes the effects of diffusion limitation of the lactose to the enzyme. The cellulose triacetate is highly stable and does not present any toxicological problems. The high operational stability of the enzyme in this process is evident from the fact that 0.5 kg of enzyme fibres (4.5×10^4 units of enzyme) operating at 4–7 °C processed 10,000 litres of milk in 50 batches with a loss of only 10% of the enzyme's activity. The other major problem, that of microbial contamination, was overcome by periodically flushing the column with bactericidal compounds.

To summarize, immobilized enzymes are (potentially) able to provide the dairy industry with processes for the continuous coagulation of milk for cheese production, turning the waste whey into an economically valuable product and rendering milk a suitable food for a large proportion of the world's population to whom it is otherwise toxic. There is one other area where immobilized enzyme technology may also be of value and that is in the stabilization of milk. Milk treated with trypsin is less susceptible to oxidation and loss of flavour, to a degree dependent upon the extent of proteolysis. The shelf life of milk has been increased by 2–3 weeks in a pilot experiment by treating milk with immobilized trypsin. The economic value of this extended shelf life with no change in flavour, in contrast to present-day sterilized and UHT milk, is potentially enormous.

e. Antibiotic Modification

So far we have dealt with processes whose economic value is based on the production of large quantities of low cost product. At the other end of the scale there are those products that are of high cost but required in relatively small quantities, for example antibiotics. The prime examples are the semisynthetic penicillins. These are based on 6-aminopenicillanic acid, the deacylated form of benzyl penicillin.

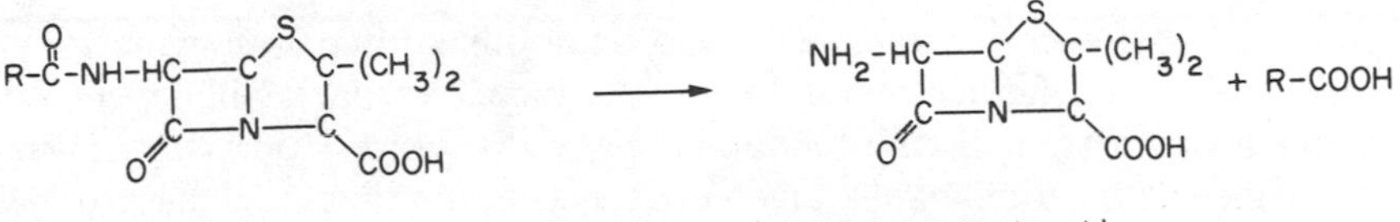

An enzyme penicillin amidase which can effect this deacylation is formed in a variety of bacteria, for example *Escherichia coli*. A number of different industrial processes based on immobilized penicillin amidase are at present in operation using, for example, enzyme immobilized with cyanogen bromide to Sephadex

G200, entrapped in cellulose triacetate fibres, in batch and packed bed reactors. More recently whole *E.coli* cells, immobilized by entrapment in polyacrylamide gel, have been used to catalyse the same reaction. The practice of using whole or lysed immobilized microbial cells or spores as catalysts is gaining in popularity, the reasons for which will be discussed later.

f. Proteases

The potential for the utilization of proteases in the food industry, other than those applications that have already been discussed, is large. The interest in this area stems from the desire to render food proteins soluble, to texturize proteins or to increase their digestibility. Three reasons exist for the use of immoblized enzymes in such processes. The first is that the use of relatively few proteases is allowed in the food industry and the potential for extending the range of processes used by immobilizing those enzymes at present banned would have many advantages. Second, many proteases are expensive and thus their re-use would reduce operating costs. The third reason is that most proteases become inactivated by autolysis, which can be prevented by immobilization.

The consumer of beer demands that his purchase must not only taste right but must also look right, that is it must be completely clear. Proteins present in untreated beer will cause a haze to develop when the beer is chilled; much time and effort is spent in the brewing industry in order to remove this chill haze. A process has recently been described whereby papain immobilized on to chitin has been used to remove the chill haze from beer. The concept of using a protease in this context is not particularly novel, but the choice of polymer support is. The low cost of chitin in comparison to most other polymers makes its use economically possible but, perhaps more importantly, chitin is a poor adsorbent (unlike for example cellulose) and thus is unlikely to remove any of the flavours from the beer.

A considerable problem in the operation of immobilized enzyme reactors, particularly the packed bed variety, is that of microbial contamination. Various solutions to this have been adopted, among them sterilizing the substrate flow or periodically flushing the reactor with a bactericidal agent, a symptomatic treatment rather than a cure. To overcome this various attempts have been made to construct a self-sterilizing packed bed reactor. The simplest technique to be described involves the co-immobilization of lysozyme with the required enzyme; whether or not it is an economic approach will depend upon the particular system in use. Another elegant solution to the problem of microbial contamination in milk involves the use of immobilized β-galactosidase/glucose oxidase for the activation of lactoperoxidase.

Lastly in this section we should mention an ingenious system of wide applicability. Many industrial and experimental processes rely upon the use of pressurized ultrafiltration membranes to retain soluble proteins within the system. All too often the protein becomes polarized around the membrane which may become clogged and is then likely to be ruptured. Velicangil and his colleagues have demonstrated a novel solution to this problem in which they immobilized various proteases to ultrafiltration membranes, with the result that the membranes were effectively self-cleaning.

g. Amino Acid Production

The biological synthesis of amino acids on an industrial scale will probably never be as quantitatively significant as some of the really large-scale processes, for example the production of high fructose syrups. Nevertheless amino acids are an important product and are likely to become more so with their increasing use as food additives. Traditional biological synthesis of amino acids has been based on microbial broth fermentation with all its associated problems of purification and decontamination of the product. The impetus towards the use of immobilized enzymes in this field, as in others, arose from the desire to save production costs by employing re-usable catalysts and more automation. One of the more interesting aspects of this area is that it is almost completely dominated by the development of the use of immobilized whole microbial cells rather than immobilized purified enzymes; the relative advantages of each are discussed at the end of this section. First, however, we shall look at some examples of amino acid production using immobilized catalysts either patented or presently in commercial use.

L-lysine. L-Lysine may be produced by the hydrolysis of DL-α-amino-ε-caprolactam which may be easily synthesized from cyclohexene, a by-product of the manufacture of nylon.

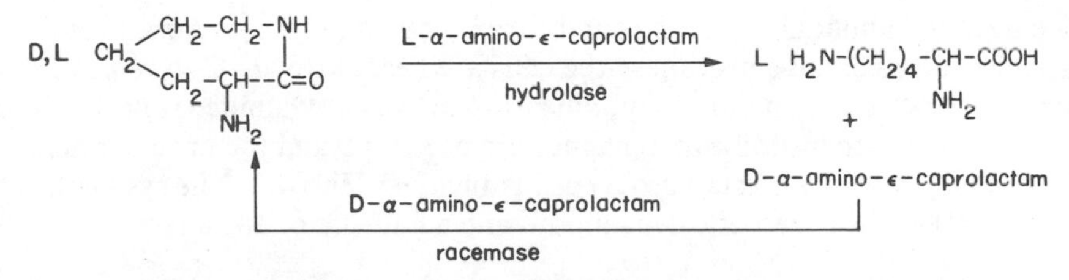

Two enzymes are involved in the process L-α-amino-ε-caprolactam hydrolase which converts the DL-α-amino-ε-caprolactam to L-lysine and the D-α-amino-ε-caprolactam; the latter is then converted enzymically to the DL form. A process using both enzymes immobilized on to an ion-exchange polysaccharide was patented by Toray Industries Inc. of Japan in 1974.

L-aspartic acid. The commercial production of L-aspartate from ammonium fumarate by immobilized *Escherichia coli* has been in existence since 1973. *E.coli* cells are harvested and immobilized by entrapment within a polyacrylamide gel and placed in a packed bed reactor. The activity of the column was found to rise dramatically over the first 24 hours of use, probably because autolysis of the trapped *E.coli* allowed easier access of the fumarate to the intracellular enzyme. The half-life of such a reactor, that is the time taken for the product yield to decrease to 50% of the original level, is typically 120 days. The long operating time of the reactor together with savings in labour costs resulting from the increased potential for automation is the major economic factor, since it markedly reduces the cost of the catalyst.

The cost of the aspartate produced by this method is some 40% less than that produced by the traditional microbial fermentation process.

L-citrulline. Pseudomonas pudita has been used, immobilized in polyacrylamide gel, in a packed bed reactor, to convert arginine to citrulline with a yield of 98% of the theoretical figure and an operational half-life of 140 days.

Urocanic acid. Unlike most other amino acids, urocanic acid has a very specific commercial application as an ultraviolet filter in suntan preparations. It is produced from L-histidine by the action of histidine ammonia lyase.

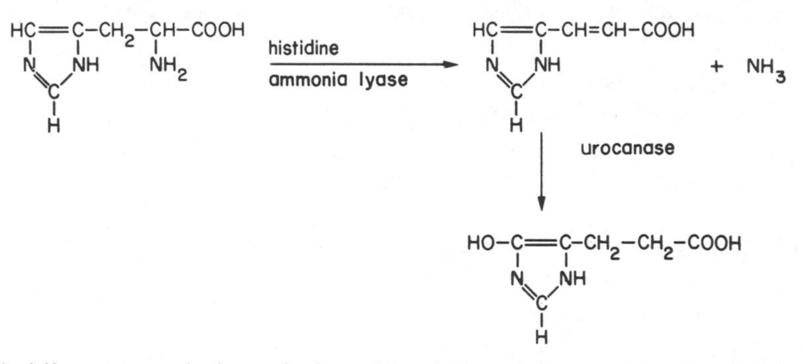

The histidine ammonia lyase is found in *Achromobacter liquidum*. Since this organism also contains urocanase the cells are heated to 70 °C to remove this latter enzyme activity before immobilization in polyacrylamide. A packed bed reactor is fed with histidine and magnesium at pH 9.0 and the urocanic acid in the product stream is crystallized by adjusting the pH to 4.7. The system has a yield of 91% of the theoretical maximum and a half-life of 180 days.

h. Immobilized Enzymes or Cells?

In recent years the practice of using non-growing whole or lysed microbial cells, rather than purified enzymes, is gaining in popularity and it is therefore pertinent to enquire as to the relative advantages of these two types of catalyst. Throughout the following discussion it must be remembered that the overriding consideration is usually one of cost.

Despite the early promises that immobilization enhanced the thermal stability of an enzyme, in fact this is rarely the case (see Chapter 2, Section II.2). In general, enzymes would appear to show little difference in their thermal stability whether *in vivo*, isolated in dilute solution, or immobilized.

In practice there is little to choose between the immobilized cell or immobilized purified enzyme as far as stability is concerned.

One theoretical advantage in the use of the immobilized purified enzyme, as opposed to the immobilized cell, is that it is possible to obtain higher loadings of catalyst. The maximum activity per gram of solid of an immobilized enzyme preparation is often ten times that of a preparation of immobilized cells from which the enzyme was obtained, although 'activation' of immobilized cells (e.g. lysis) can provide activity loading nearly as great as that of an immobilized purified enzyme. However, as immobilized catalysts of high activity are usually subjected to diffusion limitation, thus resulting in inefficient use of the catalyst, high activity catalysts are not necessarily as advantageous as they might at first seem. Clearly the extra cost involved in the purification of an enzyme will not be economic if the resulting immobilized enzyme is not used at 100% efficiency. Far better to immobilized the original cell from which the enzyme was obtained.

Although most commercially important enzymes can be derived from microbial cells some of them, notably those with a polymerase action, are usually extracellular enzymes. In such cases it would obviously be inappropriate to immobilize the microbial cell. In any event the use of immobilized cells would not be a viable proposition where the substrate is a polymer or is insoluble, because of the restriction of contact between catalyst and substrate. For that matter immobilized purified enzymes cannot be successfully employed as catalysts where the substrate is a particulate or a colloidal substance for much the same reason, unless the enzyme is immobilized on to a soluble polymer matrix. This in itself creates all manner of problems and prohibits the use of the very reactor configurations, i.e. the packed bed reactor, which may be fully automated with all the consequent economic benefits.

In many cases the cost or practical problem of isolation and purification of an enzyme was often an insurmountable factor. With the introduction of automated, continuous liquid/liquid phase separations for the isolation of intracellular microbial enzymes (see p.100), it is now possible to prepare such enzymes on the scale required for their use as industrial catalysts. Whether the effort involved in the purification is justified may depend upon the other enzymic activities demonstrated by the microbial cell. If these spurious enzymic activities do not affect the substrate or product materials, there may be little point in using the purified subject enzyme rather than the whole cell. In cases, such as the production of urocanic acid by *Achromobacter liquidum* quoted above, where the whole cell contains an enzyme that affects adversely either substrate or product, then it may be possible to remove this undesirable enzymic activity by, for example, the use of inhibitors or, more simply, selective thermal denaturation. When spurious undesirable enzyme activity cannot be eliminated there will be no alternative to the isolation and purification of the desired enzyme with its attendant cost.

Although the potential for the use of immobilized enzymes with lyase activity is by no means exhausted, future developments will almost certainly require the use of other classes of enzyme reaction, most particularly those involving the use of coenzymes. While it is theoretically possible to construct immobilized enzyme reactors to which the required coenzyme may be added with the substrate, such a system is likely to be prohibitively expensive unless the coenzyme is recycled. As coenzymes are relatively small soluble molecules which cannot be retained in the enzyme reactor as such, their extraction and regeneration are likely to be uneconomic. One solution to this problem is to immobilized the coenzyme on to a soluble polymer (e.g. dextran). In this form it may be retained within the reactor either by an ultrafilter fitted to a CST reactor or by entrapping it within another polymer alongside the enzyme. Experiments with soluble immobilized coenzymes have shown that they may be enzymically utilized but with a much reduced efficiency. Two problems still remain however, the *in situ* recycling of the coenzyme, which we shall deal with in Section V, and the instability of immobilized coenzyme. Unlike the multiple bonding that usually exists between an immobilized enzyme and its polymer support, it is not usually feasible to immobilize a coenzyme by more than one point of attachment because of their small size. This, together with the inherent instability of many coenzymes, usually means that the immobilized coenzyme will not have a sufficiently long half-life to be of any use in an industrial enzyme reactor.

In view of the problems outlined above it may be in this area that immobilized cells will be of unique value. The whole microbial cell is not only a convenient bag in which all the required coenzymes are contained, but it also contains all the necessary enzymes for their recycling. It is most likely therefore that the true potential of immobilized cells as catalysts is yet to be realized.

IV ECONOMIC AND PRACTICAL ASPECTS

In this section we shall consider some of the aspects which affect the choice of reactor and catalyst in industrial processes. We shall begin with a consideration of all the variant factors which may limit the possible reactor/catalyst configurations, continue with a discussion of the advantages of particular types of reactor and catalysts, and conclude with a detailed description of the first industrial process to utilize immobilized enzymes, the resolution of DL-amino acids.

1. Factors Affecting Choice

The most critical factor affecting the adoption of an immobilized enzyme process is one of cost. All that is discussed hereafter is largely concerned with reducing process costs without affecting product yield or safety requirements.

a. The Enzyme

The first and most obvious consideration is the nature of the conversion required: is it possible to achieve this enzymically? If so, is the enzyme readily available at moderate cost, high cost, or must it be extracted from an organism 'on site'?

If the enzyme is readily available at a low cost (in comparison to the market value of the product), then the expense of immobilization may not be worth while and it may be more economic to use the enzyme in its soluble form in a batch reactor. Whether this latter possibility is feasible will depend upon the purity and yield of product required. Where a high purity product is needed, it may be more expensive to separate soluble enzyme from the product than to immobilize the enzyme in the first instant.

Where the enzyme is expensive then the extra cost of immobilization, with its attendant promise of catalyst re-use, may be less than the cost of continually adding more enzyme to the reaction system. Thus the criterion of re-use of enzyme is important; so too is the stability of the immobilized enzyme in relation to the average residence time of substrate within the reactor. In order to obtain a high yield of product a balance must be struck between flow rate, reactor volume, and enzymic activity. High flow rates and small reactor volumes will give a small residence time and thus, for high product yield, will require a highly active enzyme preparation. Increasing the reactor volume will increase the residence time but will, at least in the case of the packed bed reactor, increase the quantity of immobilized enzyme required and thus the cost of the process. On this basis it might seem sensible to use small reactors with a low flow rate, but this will reduce the total productive capacity of the system. The alternative, of loading a polymer matrix with large quantities of highly active enzyme is not necessarily advantageous, because diffusional limitations may become a significant factor resulting in inefficient use of the enzyme (see Figure 42). In fact efficiency of use of the enzyme is the key factor

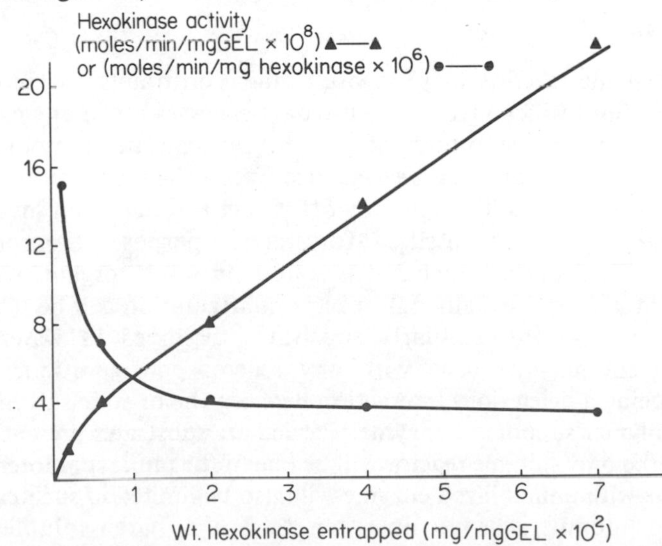

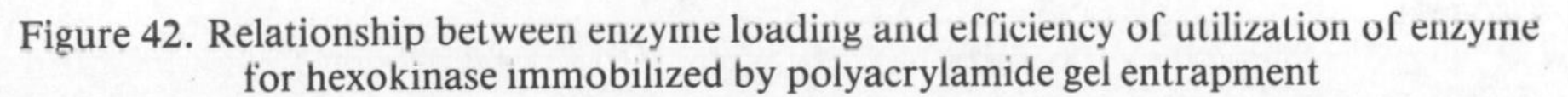
Figure 42. Relationship between enzyme loading and efficiency of utilization of enzyme for hexokinase immobilized by polyacrylamide gel entrapment

and ideally the immobilized enzyme particle must have the maximum loading of enzyme possible that will not induce significant diffusion limitation. The presence of product or substrate inhibition will complicate the matter even more (see p.87). Another consideration must be that of the operational half-life of the system; the potential for re-use of the catalyst cannot be realized unless the immobilized enzyme is sufficiently stable under operational conditions, because if the immobilized enzyme must be continually replaced, little will be gained by the use of an immobilized enzyme process, except an increase in cost.

Part of the cost of an immobilized enzyme will be contained in the polymer matrix used. It will be of great economic benefit if the polymer matrix can be recycled once its enzymic activity has been lost, particularly if the enzyme's operational half-life is rather short. It is precisely because their enzymic activity may be easily regenerated that enzymes immobilized by ionic attachment to a polymer are often the catalyst of choice, despite the potential problem of detachment of enzyme from polymer matrix.

Occasionally it may be necessary to produce the enzyme 'on site'. If the enzyme is an extracellular enzyme then the isolated enzyme must be used, not necessarily in a highly purified form. If an intracellular enzyme is required, it may make more economic sense to immobilize the whole, non-growing cell rather than incur the cost of isolation and purification of the enzyme. However, as the only practicable method of whole cell immobilization is some form of polymer entrapment and as such methods are not conducive to catalyst regeneration, a balance must be struck between the expense of enzyme isolation and purification and the potential savings permitted by catalyst regeneration.

b. The Reaction

The nature of the reaction may impose certain constraints on the choice of process. Reactions which involve the use of coenzymes are in any event liable to prove too expensive, unless the coenzyme can be immobilized and regenerated. Much research has been carried out on this topic, but its practical application is probably still a long way off. Other reactions, not involving the use of coenzymes, may be classified, for practical purposes, according to the nature of their substrate, whether it is soluble in water, organic solvents or particulate in nature. Virtually any reactor configuration can be used with a water-soluble substrate. Similarly substrates dissolved in water-miscible organic solvent may be used with any reactor configuration, the only restriction being a deleterious interaction between the organic solvent and the enzyme's polymer support or enzyme. Particulate substrates present a special problem as the only suitable reactor will be one of the tank type. Interaction of the substrate with immobilized enzyme will also be limited in such cases, thus the most economic prospect is often to use a batch-soluble enzyme process.

c. The Polymer Support

Ideally the polymer material used to immobilize the enzyme must be inexpensive, retainable within the reactor, mechanically very stable, incompressible, allow no leaching of the enzyme, be resistant to microbial attack, exhibit the desired degree of hydrophilicity/phobicity, have a high surface area to volume ratio, be capable of binding large quantities of enzyme as necessary, and exist in the desired physical form. Although some 20 years of international research has been conducted into the search for the ideal polymer, every material used is the result of some compromise.

The ability to retain the polymer within the reactor is of paramount importance. The most convenient solution is to use a packed bed reactor, but where the use of continuous flow stirred tank reactor is indicated, particle size and density become important. There are essentially three ways in which the immobilized enzyme particles may be retained within a tank reactor. If the density of the particles is sufficiently high then a separate settling tank located in the product stream may be employed. However, many support materials have a density only a little greater than one. Although there are high density polymers such as porous metals and glass their use may not be appropriate for other reasons. Thus where low density polymers must be used they must be retained by placing a filter over the reactor outlet. This may lead to polarization of the polymer particles around the filter with the result that it becomes clogged. Prepared polymer particles usually come in a range of sizes and unless the smaller particles are removed prior to use, an expensive procedure, they may pass through the filter. A compromise may be achieved by using a heterogeneous polymer with a low density periphery and a high density core made, for instance, by including metal inside a resin particle. If the metal is magnetic it may be retained by applying a magnetic field to the reactor, or made to settle rapidly in a magnetic field. An additional reason for using easily recoverable polymer particles is that the reaction mixture may contain insoluble solids which must at some stage be separated from the enzyme.

Let us first examine the possibilities for compromise in the mechanical properties of the polymer support, stability and compressibility. The compressibility of the polymer will be of little concern in a stirred tank reactor, but if a highly compressible polymer material is used in a packed bed reactor, considerable operational difficulties may be experienced. Even if the column does not become completely clogged, localized volumes within the column may become compressed. This will result in a channelling of the substrate flow, with a consequent reduction in the effective mixing of substrate and enzyme. The development of inorganic support materials, such as porous glass, was the result of a search to find alternatives to the rather compressible cellulose polymers. The mechanical stability of the polymer particle, that is its ability to withstand attrition, is likely to be a far more important factor than compressibility if the use of a stirred tank, rather than packed bed, reactor is contemplated. However, if other factors dictate the use of a continuous flow stirred tank reactor and there is significant attrition of the polymer, it may be possible to retain the degraded pieces by placing a filter over the product outlet, except in the case of a polymer-entrapped enzyme.

Leaching of enzyme can occur even where the enzyme has been covalently bound to the polymer, as certain types of bonding, for example cyanogen-bromide-activated cellulose, are in practice not 100% stable. More usually, leaching is a problem where the enzyme has been immobilized by ionic interaction with the polymer. In these cases environmental factors such as pH and ionic strength must be carefully controlled in order to minimize any leaching and this may in itself determine the appropriate reactor configuration.

In certain instances the instability of the enzyme–polymer bond may be put to good use (see the description of the L-amino acid acylase process at the end of this section), however if undesirable leaching occurs in a reactor two problems will be experienced, loss of activity of immobilized enzyme and contamination of the product stream. While the former problem may not be important if the enzyme is cheap and high loadings of enzyme on to polymer can be achieved, the latter problem may necessitate an expensive cure.

The converse of loss of material from an immobilized enzyme reactor is the addition of undesirable material, namely microbes. There are essentially two problems which may be caused by microbial contamination, spoilage of the product and, in packed bed reactors, blockage of the reactor.

The degree of risk of microbial contamination will of course depend on the process. Where a certain amount of contamination can be tolerated then, if the time taken for the contamination to rise above the maximum acceptable level is greater than the life time of the enzyme in the reactor, decontamination can be carried out when the reactor is recharged with enzyme. If, however, contamination reaches a serious level during the life time of the enzyme, the process must be interrupted and the reactor flushed with a bactericidal agent. Ideally the solution to this problem is to run the reactor as a sterile, sealed unit. Where this is not possible the risk of contamination may be minimized in a number of ways. If a material such as milk is involved the risks will be high, whereas they may be negligible if a chemically synthesized antibiotic is the starting material. No matter what the polymer support material, the very fact that the enzyme is itself a protein makes the catalyst susceptible to microbial attack. However, as an immobilized enzyme preparation consists largely of the polymer support material the problem of microbial contamination can be reduced considerably if the support is itself not a substrate for microbial growth. Thus materials such as vinyl or acrylamide polymers rather than cellulose-based polymers may be used, better still, inorganic materials such as porous glass. It is possible to entrap or bind the enzyme within the pores of the polymer support thus rendering it inaccessible to microbial attack.

Microbial contamination can be limited by adjusting the operational parameters of the reactor so that the process is carried out at high temperatures and at a pH as distant from neutrality as is possible. In this respect the ability of some polymer matrices to alter the operational pH optimum of their immobilized enzyme may be an advantage (see Chapter 2, Section III).

The polymer support material must also possess the desired degree of hydrophobicity or hydrophilicity. In general the greater degree of hydrophilicity of the support, the greater the degree of hydration and the greater the attached enzyme's activity. Certain hydrophilic polymers may, however, exist as potentially compressible swollen gels and hence will be unsuitable for use in certain reactors. Furthermore, their degree of hydration and thus structure may be affected by the presence of organic solvents in the reaction mixture. This latter consideration will be particularly important in the future application of immobilized enzymes to the processing of hydrocarbon materials such as steroids.

The surface area to volume ratio of the polymer will dictate both how much enzyme may be bound and how accessible the enzyme will be to its substrate. Small particles with a highly porous structure will have a greater surface area to volume ratio than large impervious particles, but again a balance must be struck between this parameter and the other requirements of the system. For example, the mechanical stability of a polymer falls with a reduction in particle size. It will be apparent that sacrificing polymer stability in order to obtain a large surface area and hence high enzyme loading, will be an expensive disadvantage if, at the level of loading achieved, the enzyme is inefficiently utilized because of the presence of diffusional limitations. The exception may be the case of the expensive polymer and very cheap but relatively unstable enzyme, when the presence of diffusional limitations may extend the operational half-life of the immobilized enzyme, thus reducing the process cost.

In this context general consideration must be given to the potential effect that the polymer may exert on the apparent kinetics of the reaction, in particular alteration of K_m or K_i values and the pH optimum of the enzyme (see Chapter 2, Sections III and IV).

The immobilized enzyme may be required in a peculiar physical form. While almost any polymer material can be rendered into a particulate form, only certain materials can be formed into membranes or filaments. There must also exist a consideration of safety in as much as the use of a potentially toxic polymer material will be hazardous if the product of the process is required for human or animal consumption. Last, but by no means least, the cost of both the polymer and the immobilization procedure must be considered.

2. Comparison of Reactors

When attempting to discuss the most appropriate reactor configuration for a particular process the possible interaction between the reaction and the reactor must be considered. It must be remembered, however, that a frequent and often overriding factor will probably be the nature of any existing plant or reactor equipment. For example the use of an existing continuous flow stirred tank reactor may be more economical than investing large sums in fabricating a packed bed reactor, even though the latter may be cheaper to run.

Where a new reactor must be fabricated the cost will in part be related to size and, as the size will depend upon the reaction rate available per unit volume of solvent, packed bed reactors with their low void volume (as little as 30%) are usually smaller than continuous flow stirred tank reactors which rarely have a void volume of less than 90%.

Before we begin it must be noted that all enzyme reactions must be carried out at well controlled pH and temperature. While the pH of a laboratory-scale reactor may be controlled by the use of buffers, this approach would be prohibitively expensive for use in a large-scale reactor. Thus pH must be controlled by the monitored addition of acid or alkali. In very large reactors the problem of long mixing times may make it impossible to achieve a homogeneous pH throughout the reactor, unless high stirring speeds (tank reactor) are used. This may create more of a problem than it solves, resulting in polymer particle attrition in the tank reactor or channelling in the packed bed reactor.

It will be instructive to begin by comparing directly certain aspects of two major types of reactor, the packed bed and continuous flow stirred tank reactor, as this will illustrate the nature of the factors that must be taken into account. We can then briefly discuss the relative advantages of these and other reactors.

Mathematical calculations of reactor performance can be derived from a combination of the reactor's parameters and Michaelis–Menten kinetics. Under ideal conditions such calculations can predict the quantity of enzyme required for a defined product yield or fractional substrate conversion. If we assume that the rates of substrate flow and reactor residence time are the same in each case then for the desired product yield the quantity of enzyme required will depend on several factors.

In the tank reactor all substrates and products will be homogeneously distributed throughout the reactor, while in the packed bed column reactor the substrate concentration will be at a maximum at the inlet end of the column and, conversely, the product concentration will be at a maximum at the outlet end of the column.

The effect of the relationship between substrate concentration and the K_m of the enzyme will be different in each reactor type. When the substrate concentration is much greater than the K_m the amount of enzyme activity required will be the same for both reactor configurations, as variation in substrate concentration will not affect the rate of reaction. However, if the substrate concentration is much less than the K_m the reaction will be operating under pseudo first order kinetic conditions, and the enzyme activity required in a packed bed reactor will be a fraction of that required in a continuous flow stirred tank reactor. The precise ratio of the activities will depend upon the desired degree of substrate conversion, at 90% conversion the ratio will be approximately 1 : 4 and at 99% conversion 1 : 25. This is to be expected

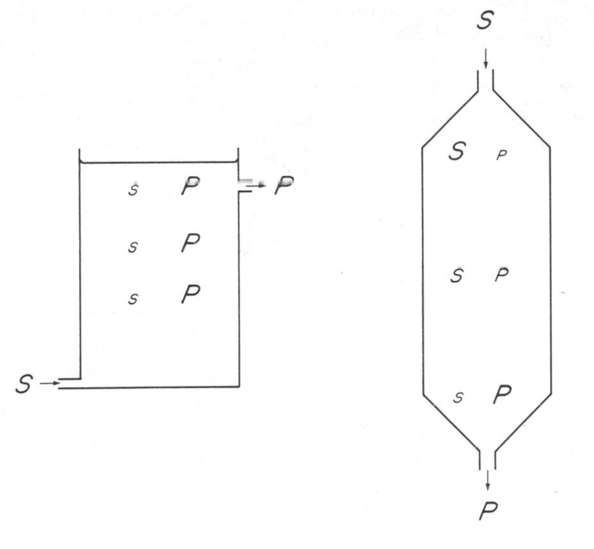

Figure 43. Substrate and product concentrations in packed bed and continuous flow stirred tank reactors. Substrate concentration $S > s$, product concentration $P > p$

because in the continuous flow stirred tank reactor the average substrate concentration must be the same as that in the outlet stream and will thus be markedly less than the substrate concentration in the inlet stream. In contrast, in the packed bed reactor the substrate concentration only approaches that of the product stream at the outlet point. Elsewhere in the reactor the substrate concentration and hence the rate of the reaction is higher, the value approaching that of the substrate stream at the reactor inlet point (see Figure 43).

The presence of product inhibition will amplify this difference between the reactors, because the product will exist in a greater average concentration in the continuous stirred tank reactor than in the packed bed reactor. It must also be noted that product inhibition is often competitive with respect to the substrate and the substrate concentration in the tank reactor is minimal. Conversely, where substrate inhibition is present the continuous flow stirred tank reactor will be the more efficient configuration. The effect of product inhibition will be lessened and the effect of substrate inhibition increased as the conversion factor is lowered or the substrate concentration of the inlet stream raised. For the mathematically minded the equations describing reactor performance under these conditions are given in Table 4.

We shall now turn to a description of the major operational advantages and disadvantages of these and other reactor configurations.

Table 4 Reactor performance equations for plug flow (PFR) and continuous flow stirred tank reactors (CSTR). Reproduced by permission from Veith *et al.*, *Applied Biochem. Bioengng*, Vol. I, p. 240 (1976)

	Reactor Performance $k_2 \cdot E_o \cdot t =$	
Reaction Conditions	CSTR	PFR
No inhibition	$S_s \cdot X + K_v[X/(1-X)]$	$S_s \cdot X - K_v \cdot \ln(1-X)$
Product inhibition (competitive)	$S_s \cdot X + K_v[X/(1-X)] + (K_v \cdot S_s \cdot X^2)/K_p(1-X)$	$S_s \cdot X(1-K_v/K_p) - K_v \cdot \ln(1-X)$
Substrate inhibition	$S_s \cdot X + K_v[X/(1-X)] + S_s^2(X-X^2)/K_s$	$S_s \cdot X - K_v \cdot \ln(1-X) + S_s^2(2X-X^2)/2K_s$

k_2 = catalytic rate constant
E_o = reactor enzyme concentration
t = reactor residence time
K_v = effective K_m of the immobilized enzyme
K_s = Substrate inhibition constant
X = fractional substrate conversion $(S_s - S_p)/S_s$
S_s = inlet substrate concentration
S_p = outlet substrate concentration
K_p = product inhibition constant

a. Continuous Flow Stirred Tank Reactors

This type of reactor is probably the least expensive to construct, or may already exist in an industrial plant. In addition it does have the advantage of ease of access, thus pH control may be effected with ease and enzyme replacement is simple. Where then the operational half-life of the immobilized enzyme is relatively short, this may be the reactor of choice. Not only is access straightforward, but effective mixing of enzyme and substrate can be obtained, thus reducing diffusion limitation to a minimum, even with low substrate flow rates. The lack of impediment to substrate flow and high mixing efficiency also means that this type of reactor may be used with colloidal or insoluble substrates, given an appropriately immobilized enzyme. The compressibility of the immobilized enzyme is of little importance, but the mechanical stability of the polymer support is important, as it will be subjected to considerable buffeting by the stirring mechanism. The limitations of use of this reactor at low substrate input concentrations and in the presence of high product concentrations have been discussed as have its advantages where substrate inhibition is present. In general it is unsuitable where high conversion reactions are required. When used with a soluble immobilized enzyme the reactor may be conveniently fitted with an ultrafiltration membrane over the outlet. However, such reactors show poor long-term stability largely due to adsorption of the enzyme on to the membrane. Furthermore their theoretical catalytic activity may be reduced because of the tendency for the enzyme to concentrate at the membrane surface.

Despite their ease of fabrication and relative ease of environmental control, tank reactors are relatively difficult to automate but have the advantage of wide applicability and ease of adaption to many different processes.

b. Packed Bed Reactors

The major disadvantage of the packed bed reactor is that it is expensive to fabricate and relatively difficult to commission. Other operational problems exist because access to the reactor is limited. Thus, this type of reactor is difficult to recharge with immobilized enzyme and so it is not ideally suited for use with enzymes with a short operational half-life. In addition, it may be difficult to control the pH of the whole column, particularly if the reaction is acid or base liberating. However, the packed bed reactor does have considerable advantages which usually outweigh its inconveniences. For example, as we have discussed above, it is capable of high substrate conversions, even in the presence of product inhibition, using less enzyme than a tank reactor. Perhaps most important of all is the relative ease with which it may be automated, this being one of the major cost benefits of this type of reactor. Thus, once operational, the reactor will virtually look after itself, although certain factors may cause premature ageing. Bacterial contamination may be a problem possibly causing blockage or disturbance of flow through the reactor. Another problem may arise from the use of a compressible polymer support, that of localized channelling of flow which will lead to an effective reduction of the operational half-life of the reactor. Here in particular it is important to consider the interaction between the polymer support and the viscosity of the substrate stream. While a low viscosity substrate stream may cause no problems with a particular polymer support material, the extra flow pressure required to drive a high viscosity fluid through the reactor may cause channelling in the immobilized enzyme bed. Two solutions exist to this problem, reduction of substrate viscosity or redesign of the reactor. If the substrate is, for example, a viscous starch solution it may be possible to incorporate a predigestion process, using a continuous stirred tank reactor, before passing the substrate stream into the packed bed reactor. Alternatively, the reactor may be designed as a flat bed reactor thus reducing the pressure drop across the reactor. Channelling may also be caused by the formation of air bubbles within the reactor bed. Prewarming the substrate stream and placing the inlet at the top of the column may help eliminate this hazard.

Packed bed reactors are much less sensitive to polymer attrition than continuous flow stirred tank reactors and contamination of the product stream with polymer particles is much less likely to occur.

Finally it must be realized that although choice of reactor can be made on theoretical grounds, such considerations are based on the assumption that the reactor behaves as an ideal plug flow reactor (i.e. there is no variation of solvent flow rate across the column). When the reactor deviates from this ideal pattern of behaviour there will occur substantial falls in reactor efficiency.

c. Fluidized Bed Reactors

A fluidized bed reactor is in theory blessed with all the advantages of the packed bed reactor together with advantages peculiar to itself. Where a highly exothermic reaction is concerned, heat transfer and removal from the column is more easily effected, because of the fluid flow characteristics. Not only is heat transfer improved but, by the same token, the mass transfer limitations are reduced. This has two implications. The immobilized enzyme may be fabricated to contain more enzyme activity without reducing the efficiency of use of the enzyme, thus improving productivity by allowing the use of higher flow rates and shorter reactor residence times. Second, this reactor configuration may be used with colloidal, viscous, or insoluble substrates. Fluidization of the immobilized enzyme bed also eliminates problems of plugging and channelling in the reactor. The reader might be excused for asking why, with all these advantages, is the fluidized bed reactor not used to the exclusion of other types? The answer to this question is twofold, cost and uncertainty. Fluidization of a bed requires a large input of power and, although the fluidized bed reactor works reasonably on a laboratory scale, it is difficult to predict the success of a scale-up procedure.

d. Ageing in Reactors

There are a number of causes for loss of productivity of a reactor and these are summarized below.

Productivity will be lowered if enzyme activity is lost from the reactor. This may be caused either by a reduction in the activity of enzyme in the reactor or a loss of enzyme from the reactor. Loss of activity is usually a result of a time-dependent denaturation of the enzyme, but may be caused by poisoning of the enzyme by heavy metal ions or other inhibitors present in the reaction mixture. Loss of enzyme from the reactor may either be caused by detachment of the enzyme from its polymer support, most commonly where the enzyme is attached by ionic bonding, or by direct solubilization of the polymer itself. An example of this latter is carboxylmethylcellulose where certain areas of the cellulose may be more highly substituted than others and may in time dissolve in the reaction mixture. Attrition of the polymer releases small polymer particles which may dissolve in the reaction mixture, be lost as insoluble fines, or may release an entrapped enzyme.

Reduction of contact between enzyme and substrate will also lead to a loss of reactor productivity. This may be due to any combination of three possible causes, changes in flow patterns (e.g. channelling in a packed bed reactor), heterogeneous enzyme distribution (e.g. polarization of immobilized enzyme around a filter), or fouling of the surface of the polymer particle by solids present in the reaction mixture.

The third cause of lost productivity is microbial contamination which has already been discussed.

3. The Development of L-Amino acid Acylase

We can now turn to a consideration of the development of the first commercial immobilized enzyme process. The outline of the process has been discussed earlier in this chapter (Section III.2.a), so we shall content ourselves here with a discussion of the factors and choices involved in the development of the process. The work was carried out in the laboratories of the Tanabe Seiyaku Co. of Japan mainly by two workers Ichiro Chibata and Tetsuya Tosa (see Bibliography).

a. Immobilization

The first problem to be overcome was the development of a suitable immobilized preparation of the aminoacylase. Chibata and Tosa experimented with some 40 different immobilization procedures ranging from simple physical adsorption, through covalent and ionic bonding, to chemical cross-linking and lattice entrapment. The physical adsorption methods were discarded because of the low yields and low activities of the preparations. Of the ionic binding methods the weak base derivatives of cellulose and Sephadex gave relatively high yields and activities whereas ionic synthetic resins (e.g. Amberlite) and the acidic derivatives of Sephadex did not give active preparations. Covalent bonding of the aminoacylase to arylamino-glass produced one of the most active preparations, but the method was discarded as unsuitable for industrial applications because of the instability of the carrier matrix. Out of all the other activated polymers tested only the halogenacetyl celluloses, in particular iodoacetyl cellulose, gave reasonably active preparations. Cross-linking the aminoacylase with glutaraldehyde produced an immobilized enzyme preparation of reasonably high activity (about half that of the iodoacetyl cellulose enzyme) but the total yield of activity was low, only some 15% compared to the iodoacetyl cellulose derivative's 40%. The high activity of the preparation but low yield of activity, is a consequence of the absence of an inert carrier molecule in the preparation. One other successful technique was that of lattice entrapment in polyacrylamide gel.

b. Comparison of Preparations

For the second part of their study Chibata and Tosa chose the three most promising immobilization methods, ionic binding to DEAE–Sephadex, covalent bonding to iodoacetyl cellulose, and entrapment within a poly-acrylamide gel. They directly compared the parameters of these preparations with each other and the soluble aminoacylase on a laboratory scale, testing for optimal environmental conditions, changes in the apparent K_m and V_{max}, operational stability and yield. Their results are presented in Table 5 and are worthy of some comment. The pH optimum of the DEAE–Sephadex aminoacylase was found to be approximately 0.5 to 1.0 pH unit lower than that of the soluble enzyme, presumably due to the partitioning of hydrogen ions away from the positively charged polymer matrix. The decrease in optimum pH of the polyacrylamide-entrapped aminoacylase has never been explained. It is difficult to assess the significance of the figures of optimum temperature for the various enzyme systems. They may reflect genuine changes in the thermal stability of the bound enzyme but are perhaps more likely to be due to diffusional effects (see Chapter 2, Section II). That the optimum cobalt concentration is identical in all four cases and the K_m values more or less the same is surprising. It might be expected that, due to the ionic nature of the DEAE–Sephadex, this preparation might show a marked decrease in K_m (see Chapter 2, Section IV.1). It is unlikely that the ionic strength of the reaction mixture is sufficiently high to eliminate partitioning effects, otherwise the pH optimum of the DEAE–Sephadex preparation would not show the downward shift described. It is equally unlikely that a downward shift of K_m, due to partitioning, was moderated by diffusion limitation effects, because the polyacrylamide-entrapped enzyme might then be expected to show a marked increase in K_m as, of the two preparations, diffusion limitation will almost certainly be greater in the latter preparation. In practice the reason matters little, no preparation shows any marked change in K_m, thus on these grounds any preparation would be suitable for use in a reactor. The apparent increase in V_{max}, above the value of the free solution aminoacylase, for the DEAE–Sephadex and polyacrylamide-entrapped enzymes is probably a reflection of the fact that all preparations were assayed at pH 7.0, the optimum pH for the DEAE–Sephadex and polyacrylamide enzymes, but below the pH optimum of the native enzyme. The apparent tripling of the V_{max} for the iodoacetyl enzyme is a mystery. The heat stabilities of all three immobilized preparations were greater than that of the soluble enzyme, but here again this might be due to diffusional or protein concentration effects (Chapter 2, Section II). The DEAE–Sephadex aminoacylase had the longest operational half-life.

c. Choice of Catalyst Preparation

With the possible exception of the long operational half-life of the DEAE–Sephadex preparation, there is little from which to choose between the performances of the various preparations. The choice then is governed by six factors, ease of preparation, immobilized enzyme activity, cost, binding force, operational stability, and possibility of catalyst regeneration. All three preparations had high enzyme activity and could thus be considered suitable. Cost may be subdivided into two categories one depending upon the immobilization process, the easier the process the cheaper it becomes, the other depending upon the cost of the polymer support material. Whereas immobilization to DEAE–Sephadex is simple and therefore inexpensive, immobilization to iodoacetyl cellulose is complex and costly. Furthermore neither polymer is cheap, in fact in operational terms the cost of the polymer is greater than that of the enzyme, but the DEAE–Sephadex has the unique advantage that it can be recycled simply by adding more soluble enzyme to the column when the activity of the preparation falls below the desired level. Thus a potential weakness of an immobilization method, the relatively weak ionic bonding of enzyme to support, is turned into a positive advantage. It was for this reason and its high operational stability that Chibata and Tosa selected the DEAE-Sephadex aminoacylase for use in the industrial process. These factors are summarized in Table 5.

Table 5 Summary of enzymic properties of various immobilized aminoacylases. Reproduced with permission from Chibata and Tosa, *Applied Biochem. Bioengng*, Vol. I, p. 334 (1976)

Properties	Native[(a)]	Immobilized Aminoacylase[(a)] DEAE-Sephadex	Iodoacetyl Cellulose	Polyacrylamide
Optimum pH	7.5–8.0	7.0	7.5–8.0	7.0
Optimum temp.	60 °C	72 °C	55 °C	65 °C
Activation[(b)] energy (kcal/mol)	6.9	7.0	3.9	5.3
Optimum Co^{2+} (mmol l^{-1})	0.5	0.5	0.5	0.5
K_m (mmol l^{-1})[(b)]	5.7	8.7	6.7	5.0
V_{max} (mol/h)[(b)]	1.52	3.33	4.65	2.33
Heat stability[(c)]				
60 °C, 10 min (%)	62.5	100	77.5	79.5
70 °C, 10 min (%)	12.5	87.5	62.5	34.5
Operational[(d)] stability (days)	—	65 days @ 50 °C	—	48 days @ 37 °C
Preparation	—	Easy	Difficult	Medium
Enzyme activity	—	High	High	High
Cost of immobilization	—	Low	High	Moderate
Binding force	—	Medium	Strong	Strong
Operational stability	—	High	—	Moderate
Regeneration	—	Possible	Impossible	Impossible

(a) Substrate acetyl-DL-methionine
(b) Assayed at 37 °C and pH 7.0
(c) Remaining activity
(d) Time required for 50% enzyme activity to be lost

d. Operation and Economics of the Reactor

Chibata and Tosa showed that with a packed bed reactor the actual column dimensions did not affect the reaction rate recorded per column unit volume. At an operating temperature of 50 °C they showed that for 0.2 mol l^{-1} substrate, a flow rate of 2.8 space velocities (SV) per hour (i.e. a flow rate of 2.8 times the column void volume per hour) gave 100% hydrolysis of the L-acetyl methionine. Increasing the flow rate to 8 SV h^{-1} progressively decreased the percentage hydrolysis to 40%. Thus a 1,000 litre column converted 2,000 litres of 0.2 mol l^{-1} acetyl-DL-methionine per hour at 50 °C.

The activity of the column decreased gradually over a period of 30 days to 60% of the initial activity, whereupon it was regenerated to 100% by adding fresh soluble enzyme to the column. The DEAE–Sephadex proved to be a very stable polymer support, showing no loss of binding capacity or physical structure after several years' continuous use. The effluent from the column was evaporated, the L-methionine separated and recrystallized and the acetyl-D-methionine racemized by heating at 60 °C with acetic anhydride and re-used. The total yield of L-methionine was 91% of the theoretical maximum. Figure 41 shows the flow diagram of the fully automated continuous production process.

Economic analysis of the immobilized enzyme process in comparison to the conventional batch process shows a saving of nearly 50% in production costs (Figure 44). The savings are made in three major cost areas, enzyme, substrate, and labour. The cost of the enzyme in the immobilized enzyme process is understandably about one-twentieth that of the batch process. Full automation of the continuous process markedly reduced labour costs, though it must be stressed that more skill is required to operate the continuous column process. Perhaps more surprising is the saving in substrate cost, which is largely due to the simplicity of the purification procedure and higher yield of the immobilized process. The extra cost of the DEAE–Sephadex was minimal because of its long-term stability and potential for re-use. Fuel costs are higher

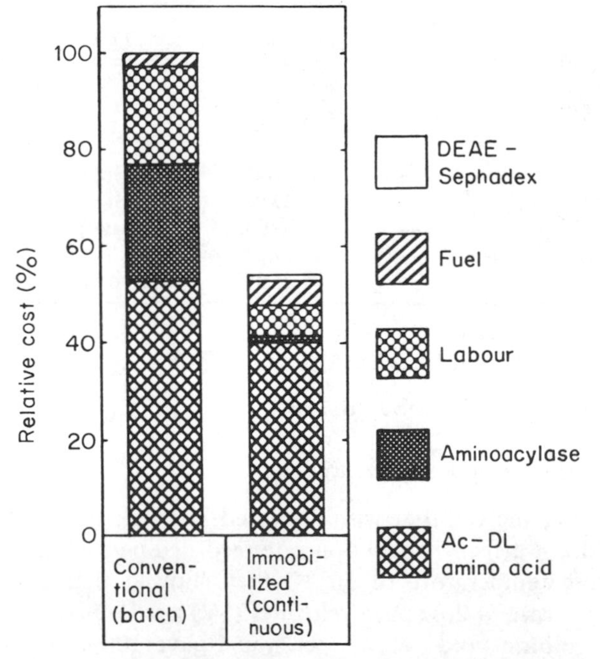

Figure 44. Economic comparison of conventional batch and immoblized aminoacylase processes. Reproduced by permission of The Society of Fermentation Technology, Japan

for the immobilized enzyme system but do not contribute more than 10% of the total cost. Neither analysis takes account of capital equipment costs.

4. Conclusion

The asiduous reader will have gathered that general conclusions concerning the choices available in the design of an immobilized enzyme process are not easily reached. However five factors are likely to be of paramount importance, the operational stability of the immobilized catalyst, the cost of the immobilized catalyst, the activity and yield of immoblized catalyst, regeneratability of the catalyst, and the cost/availability of the appropriate reactor configuration. Every process must ultimately be thoroughly analysed and tested and then considered on its merits.

V TOWARDS THE FUTURE

Attempting to predict the future is always a risky business, more so when the predictions are being committed to print, for in the period between the writing and reading, some future trends may have become past developments, while other possible developments may have been shown to be impossible. Nevertheless such prediction is a valuable pastime, if only to point the way to areas which may usefully bear investigation.

Of the many potential developments that could be discussed in this section we shall confine ourselves to five principal areas, coenzyme recycling, conversion of hydrophobic substrates, multistep synthesis, waste utilization, and processes concerned with energy production. It would appear that, at the time of writing, one of the major preoccupations of technology is the diminution of traditional energy sources, the fossil fuels. Broadly speaking the solution to this latter problem lies in either conserving energy and/or developing new sources of energy based on renewable resources. Allied to this is the problem of developing alternative precursors for those materials presently manufactured from oil and coal. The effective retention and recycling of coenzymes within a reactor are relevant to this discussion for two reasons. First, there are many chemical conversions that are at present carried out which involve oxidation/reduction reactions. The industrial chemical techniques to perform these conversions often need vast quantities of energy to supply the high pressures and temperatures required in order that the reaction may proceed. If the same conversion can be carried out enzymically normal pressures and temperatures could be used in the reactor. At present, however, most enzyme dehydrogenations (or hydrogenations) would require the continual addition of expensive coenzymes to the reactor, the cost of which far outweighs any energy saving. The second reason for developing immobilized coenzymes lies in the potential use of the oxidation/reduction reactions to which they could be applied, for example the oxidation/reduction of steroids.

1. Coenzyme Recycling and Retention

In this section we shall consider just two coenzymes, arguably the most important, NAD and ATP. The lessons that we can learn from these may be applied to other coenzymes.

Generally speaking two conflicting criteria must be satisfied when considering coenzyme retention, the coenzyme must be modified in order to retain it within the reactor but such modification must not affect its interaction with the enzyme. In addition to this the coenzyme must be regenerated, either chemically, physically, or enzymically, once it has been utilized and any reagents or apparatus for this must also be contained within the reactor.

a. NAD/NADH

Regeneration of reduced nicotinamide adenine dinucleotide (NADH), or for that matter any other oxido-reductase coenzyme, may be achieved electrochemically, chemically, or enzymically. Where the coenzyme is to be used in an enzyme reactor, enzymic regeneration is probably the method of choice. Many schemes for the enzymic recycling of NAD/NADH have been proposed and some of them are shown in Figure 45.

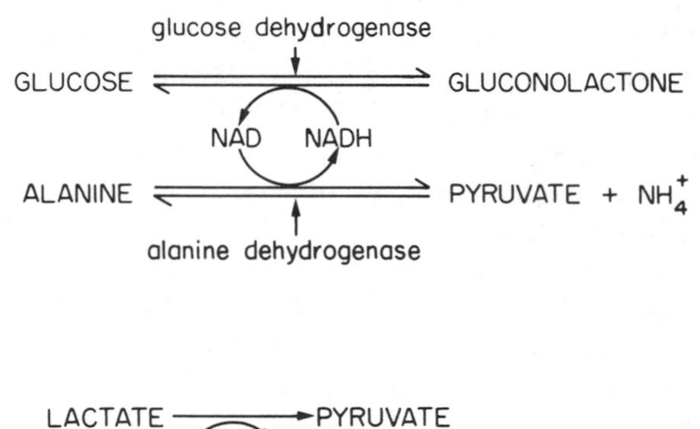

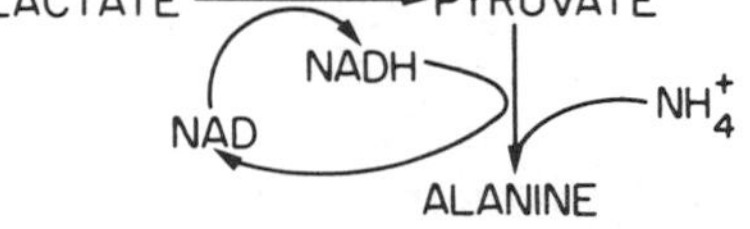

Figure 45. Reactions for NAD/NADH recycling

Where enzymic recycling is used then usually two substrates must be added to the reactor and two products must be produced, only one of which is required. Thus an additional step must be introduced to separate out the required product. Occasionally it may be possible to eliminate this complication by using two consecutive reactions, one oxidative the other reductive, as in the example of lactate to alanine given in Figure 45. Retention of NAD within the reactor may be achieved by covalently coupling it to a high molecular weight polymer. In order to allow effective contact between immobilized enzyme and immobilized coenzyme, soluble polymers, for example dextrans, polylysine, or alginic acid, are usually used. The soluble immobilized coenzyme may be retained within a continuous flow stirred tank reactor by fitting the outlet with an ultrafiltration membrane. The use of immobilized coenzymes in a packed bed reactor is more problematic and must be restricted to those situations where the enzyme is immobilized by some form of entrapment method, for example polyacrylamide gel or fibre entrapment; the enzyme and soluble immobilized coenzyme can then be coentrapped within the polymer.

Various methods have been devised for the immobilization of NAD and in general these show that direct bonding of the NAD to the polymer molecule does not give a very reactive product, greater reactivity is obtained by including a spacer molecule between the polymer and the NAD. The first step in the immobilization process is to create an NAD analogue by building the spacer molecule on to the NAD. This appears to be more successful than first building the spacer molecule on to the polymer because, after reaction of the substituted polymer with the NAD, spare spacer molecules on the polymer must be removed or covered up. Of the spacer groups tested two of the most promising are carboxymethyl and (6-aminohexyl) carbamoylmethyl groups. The point of attachment of these groups to the NAD molecule is critical, the only acceptable point being the N-6 atom. A curious difference exists between NAD analogues and NADP analogues. Whereas N-6 substituted carboxymethyl NAD and N-6 substituted (6-aminohexyl) carbamoylmethyl NAD are both equally effective coenzymes, the N-1 substituted NAD analogues show no coenzyme activity. In contrast to this both N-1 and N-6 carboxymethyl NADP are effective coenzymes, but substitution at either N-1 or N-6 with the (6-aminohexyl) carbamoylmethyl groups results in a non-reactive NADP analogue. Clearly, considerations of charge are more important in producing NADP analogues than in producing NAD analogues. The next stage in the process is to cross-link the coenzyme analogue to the appropriate soluble polymer. This may be accomplished in a number of ways depending upon the nature of the spacer entity, for example the (6-aminohexyl) carbamoylmethyl coenzyme may be reacted with cyanogen-bromide-activated dextran or Sepharose. The cyanogen bromide activation process must be carefully controlled lest the polymer becomes too highly cross-linked and is rendered insoluble. A large number of variations are possible, some with interesting potential. The selection of spacer group and polymer molecule will ultimately depend upon the enzyme with which the immobilized coenzyme is to be used, for example yeast alcohol dehydrogenase is inactivated by N-6 carboxylmethyl NAD coupled to 1,6-diaminohexane dextran, while liver alcohol dehydrogenase is not. One of the more ingenious methods involves coupling NAD to alginic acid, the immobilized coenzyme thus produced is normally soluble, but lowering the pH to below 2.5 precipitates the alginic acid, permitting easy recovery of the alginic acid immobilized NAD from the reaction mixture.

From the commercial point of view immobilized NAD is not as yet an economic proposition because it tends to be both expensive to produce and insufficiently stable. Doubtless future developments will overcome these problems.

b. ATP/ADP

ATP may be retained within a reactor in exactly the same manner as NAD or NADP, that is by substitution to produce a reactive ATP analogue followed by immobilization to a soluble polymer. Regenerating ATP in a reaction is more difficult. Many approaches have been examined in recent years with varying degrees of success. The requirements for ATP regeneration from ADP or AMP are that the phosphorylating agent should be cheap, stable, non-toxic, produce few (or no) by-products which can be easily removed, have a higher free energy of hydrolysis than ATP and the transphosphorylating enzyme must also be inexpensive. Possible phosphorylating agents include phosphoenolpyruvate, acetyl phosphate, carbamoyl phosphate, and inorganic polyphosphates. Phosphoenolpyruvate and carbamoyl phosphate can probably be ruled out on the grounds of their cost, several times that of ATP, and the cost of the respective enzymes. Inorganic polyphosphate is commercially available in bulk and is a fraction of the cost of acetyl phosphate, both of which are cheaper per mole of phosphate than ATP. However, the latter has a higher free energy of hydrolysis. The drawback lies in the cost of the enzymes, acetyl kinase is at present £5 per milligram while polyphosphate kinase is not commercially available. In the event that either enzyme could be produced inexpensively, polyphosphate does have one great advantage in that the total recycling reaction will only produce phosphate as a by-product. Unfortunately many kinases are inhibited by high phosphate concentrations, therefore recycling ATP in a stirred tank reactor may prove to be an intractable problem unless low product yields are acceptable; the use of packed bed reactors with the components retained by entrapment is a more hopeful prospect, but at present far more costly than continually adding ATP.

The outlook therefore for the development of industrial processes involving coenzymes in the reaction, in particular those involving ATP, on the basis of pure immobilized enzymes and coenzymes is not very promising. A more likely solution to the problem probably lies in the use of whole immobilized, possibly modified, cells, which contain all the necessary enzyme and coenzyme components. It would be possible to utilize non-metabolizing cells or spores if the process requires only oxidation/reduction reactions, so long as the relevant cosubstrates necessary to complete the oxido–reduction cycle are added to the reactor. Processes involving phosphorylations would require the use of metabolizing cells and the addition of an appropriate growth medium to the reaction mixture. One method of ATP regeneration that is beginning to emerge is that of photosynthetic regeneration, but as yet it is too soon to determine its large-scale potential.

2. Conversion of Hydrophobic Substrates

Many chemical processes involve hydrophobic substrates, two of the most economically important being the modification and distillation of hydrocarbons from mineral oil and the modification of steroids.

a. Modification of Mineral Oil Hydrocarbons

The fractionation and chemical modification of hydrocarbons in the processing of mineral oils is becoming increasingly expensive. This is due, in part at least, to the enormous energy requirements of a traditional oil refinery. Much research has been applied in recent years to the isolation and characterization of microbes and enzymes capable of cleaving and otherwise modifying hydrocarbon molecules. It would appear that a biologically based plant for the processing of mineral oil is a possibility. However, with the present bleak estimates of world oil resources, it would seem unlikely that the huge sums of money required to develop such a process could be recouped before the oil effectively runs out. Nevertheless the ability to modify hydrocarbons is likely to remain important long beyond the oil era, because society will still need plastics, presumably derived from the reduction of carbohydrates. At present processes exist for the production of plastic precursors from glucose but as yet are uneconomic. One major problem to be overcome is the unstable nature of hydrocarbon modifying enzymes which, together with the development of procedures for handling water-insoluble substrates in a reactor, will be subjected to much study in years to come.

b. Steroid Conversions

The past 30 years have seen the development of microbial processes for the conversion of steroids. The economic importance of this work may be judged from the fact that 30 years ago the chemical conversion of deoxycholic acid to cortisone involved 37 steps and had a yield of only 0.16%. Today, using microbial fermentations, cortisone production is far simpler and it is now a mere 0.002% the cost of 30 years ago. The isolation of steroid-converting enzymes, their immobilization and application in automated continuous reactors does not promise such startling economies, but nevertheless will undoubtedly become economically attractive. In order to enhance the performance of such reactors some method must be employed to increase the availability of water-insoluble substrates in the aqueous reactor system. To date two solutions to this problem have been proposed: the dissolution of the steroid in a silicone 'reservoir' contained in the reactor, or, alternatively, dissolving the steroid in a water-miscible solvent, for example ethyl acetate.

Various steroid-modifying reactors have been constructed on a laboratory scale but none is yet in commercial use.

The financial rewards available in this field are enormous. World production of steroids is now of the order of several thousand tonnes per year and the cost of many of the individual steroids is in the region of several pounds sterling per gram. At present there are only two major sources of substrate for steroid synthesis, diosgenin obtained from the roots of the barbasco which grows wild in Mexico and stigmasterol from soy steroids. Not only is the demand for steroids growing but the supply of barbasco is declining, thus increasing emphasis will be placed on manipulating other naturally occurring steroids, for example β-sitosterol, some 20,000 tonnes of which is produced annually (1974) in the USA as a waste product of wood pulping, or cholesterol.

3. Multistep Synthesis — Gramicidin S

While most of the present-day applications of immobilized enzymes have been limited to one-step catalysis the potential for multistep processes is large. Recent studies have shown that it is economically possible to synthesize gramicidin S *in vitro*. Such a study has been made possible in part by the development of techniques for large-scale isolation of purified intracellular microbial enzymes. Traditional protein separation techniques are not very effective in isolating such enzymes, largely because of the size similarity between the enzymes and the cell fragments left after disruption. The breakthrough came with the development of two-phase partitioning in high molecular weight polymer systems.

Two polymers, for example polyethyleneglycol and dextran, when dissolved in water at appropriate concentrations are mutually immiscible. For reasons beyond the scope of this text, certain enzymes will preferentially dissolve in one or other of the polymer solutions and thus may be separated and purified. The large-scale utilization of two-phase partitioning is further enhanced because it can be adapted into a continuous process, providing one-step separation and purification. For example, using 1.6% polyethyleneglycol 4,500 and 6.5% dextran 82,000, Wang and his colleagues achieved a one-step, 8.5-fold purification with 70% recovery of gramicidin S synthetase.

Gramicidin S is a cyclic decapeptide (Figure 46) synthesized non-ribosomally from L-leucine, L-proline, L-valine, L-ornithine, and D-or L-phenylalanine. The reaction requires the participation of ATP, Mg^{2+}, and a reducing agent. Gramicidin S synthetase is a multienzyme complex consisting of at least five different peptides. Wang and his colleagues were able to demonstrate an efficient continuous synthesis of gramicidin S in the presence of an ATP regenerating system comprised of acetyl phosphate, ATP, acetate kinase, and adenylate kinase. Their estimate showed that a process based on an immobilized enzyme should be economically feasible. The real importance of their work lies not so much in the specific production of gramicidin S, a product of little economic value, but in demonstrating the viability of the cell-free multienzyme synthesis of biological compounds.

4. Waste Utilization

Any process that can be designed to utilize as its starting material a material hitherto regarded as a waste product will have two advantages. The first is the obvious saving in cost of the substrate, the other is the saving obtained by removing the necessity to dispose of a waste product, an often expensive business. The processing of one such waste product, acid whey from cheese manufacture, has already been discussed and we shall limit ourselves here to a consideration of the utilization of cellulose waste.

Cellulose is the most abundant organic material on earth and, unlike so many other resources, is renewable. It is ironic that cellulose is the major constituent of domestic and commercial wastes and large quantities are also eschewed by the food processing industries each year. Glucose syrups obtained from cellulose would have many applications in foodstuffs, fermentations, and increasingly important, conversion to alcohol for fuel and reduction to hydrocarbons for the synthesis of plastics. In essence the problem involves the design of a process for the enzymic hydrolysis of cellulose. Unfortunately this is not as simple as it might at first appear, the nature of the cellulase enzymes and cellulose itself posing serious problems. Cellulose usually exists as a highly insoluble crystalline substance of low surface area. In addition other compounds, for example lignins, may be present which can inhibit or hinder the action of cellulases. Thus some pretreatment of the cellulose is necessary in order to make it more available to the enzyme. Although cellulase may be simply derived from a variety of micro-organisms it demonstrates a low specific activity, which in turn necessitates long reaction times. Even assuming

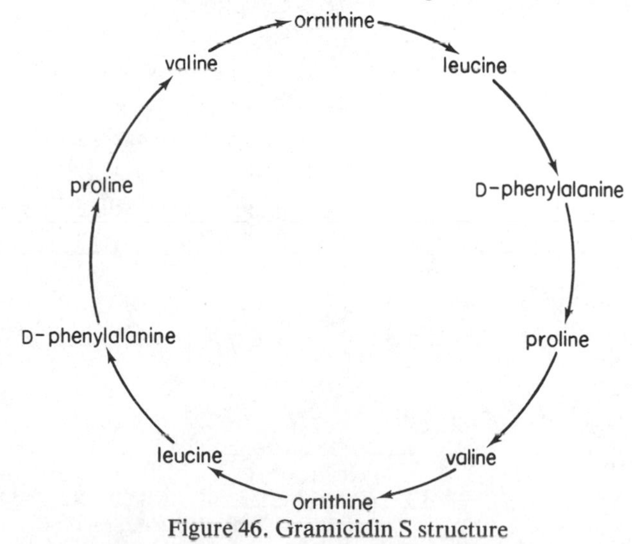

Figure 46. Gramicidin S structure

that cellulases of high specific activity can be isolated it will not be possible to adapt the cellulose process to conventional immobilized enzyme techniques because of the insolubility of cellulose. It may be possible, however, to retain the cellulase within a continuous flow stirred tank reactor by ultrafiltration which would also have the advantage of retaining the cellulose within the reactor. Alternatively it may be possible to utilize the property of cellulase to bind strongly on to its cellulose substrate, by constructing a fluidized bed reactor to which particulate cellulose is added at the top of the reactor, the cellulase being retained on the cellulose particles.

5. Energy Production

A major preoccupation of technology today is the development of alternative sources of energy. Biological reactions can be adapted to provide energy in one of two ways, either by producing electrical energy from an electrobiochemical cell or by forming a combustible product, for instance methane or hydrogen. We shall deal with each of the possibilities in turn, but it is relevant to note at this point that similarities exist in each process.

A variety of designs of biochemical fuel cell have been developed, each utilizing as the source of electrical current the electrochemical oxidation of hydrogen at a platinum black electrode (anode). The hydrogen may be produced from a carbohydrate source by micro-organisms, for example *Escherichia coli* or *Clostridium butyricum*, that contain a hydrogen-producing system based on hydrogenase. Figure 47 shows schematically one possible arrangement of such a fuel cell, developed by Karube *et al.* (1977). The anode is a platinum black electrode the surface of which is covered by a layer of, for example, *C. butyricum* immobilized by polyacrylamide gel entrapment.

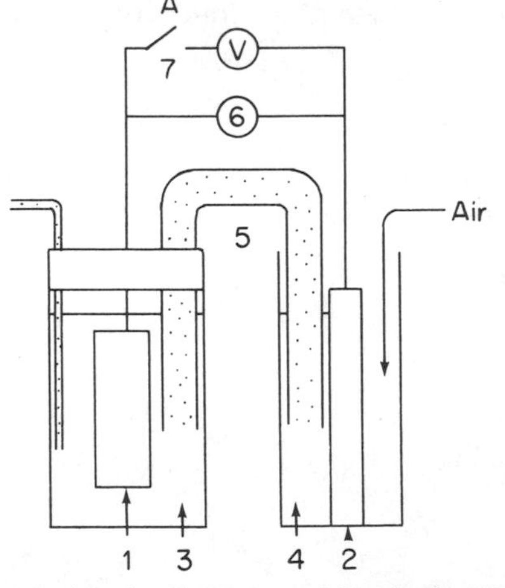

Figure 47. Diagram of biochemical fuel cell. (1) Immobilized cell (*Clostridium butyricum*) — platinum black electrode. (2) Carbon electrode. (3) Anolyte (buffered glucose). (4) Catholyte. (5) Saturated KCl bridge. (6) Recorder. (7) Switch. Reproduced from Karube *et al.* (1977) by permission of John Wiley & Sons Inc

Substrate glucose is fed continuously into the anode chamber and is converted at the surface of the anode into hydrogen and formic acid. It is possible that part of the current generated may be derived from the direct electrochemical oxidation of the glucose or formic acid. Such systems can never develop the current-generating potential of a standard chemical cell presumably because of the milder physical conditions employed in the biochemical cell. Typical potential differences of such cells are of the order of 0.6 V developing a current of 1 mA. It is therefore unlikely that such cells will ever be able to replace chemical fuel cells, however they do have the advantage of running off industrial carbohydrate wastes. Future development must be directed towards enhancing the stability of the labile hydrogenase systems.

Hydrogen-producing systems can be used *per se* to produce hydrogen as a fuel. Perhaps the most promising approach will be the utilization of the biophotolytic oxidation of water. Many organisms, for example blue-green algae and some species of bacteria can carry out this process in the presence of sunlight. Either immobilized whole cells or enzymes could be employed, but until their inherent instability has been overcome little progress will be made. If such a system is ever developed successfully it will not only bring with it great financial rewards but will provide a virtually unlimited source of pollution-free fuel, the product of combustion of which — water — unlike the products of combustion of fossil fucls, is thc substratc for thc fucl. In this sense, at a global level, the fuel becomes virtually self-renewing and must therefore be regarded as the ultimate in combustible fuel technology.

Chapter 4

Model Systems

Much has been discovered in recent years about the molecular organization of the cell. No longer is the cell regarded as a homogeneous bag of enzymes and substrates, but it is recognized that the cell organizes its constituents in a very particular way. For example its metabolic pathways may be organized more efficiently, substrates for enzyme reactions being provided at the right place at the right time. Membrane transport is another feature of cellular life, yet little is known of the role this has to play in intracellular partitioning, or the molecular nature of the membrane transference. Oscillatory phenomena, spontaneous heterogenesis (i.e. the conversion of a near 'homogeneous' system into a heterogeneous one) and hysteresis effects are now all recognized as occurring in the cell. The importance of these effects in terms of cellular control, cell differentiation, or the biochemical basis of memory is difficult to assess accurately at present. However, much can be learned from studies of immobilized enzymes set up to model biological systems. This chapter is divided into five sections dealing with the modification of enzyme reactions by mechanical stress, model multienzyme systems, models of active transport, spontaneous heterogenesis, and lastly the relevance of such studies to biological systems.

I MECHANOCATALYSIS

So far in this book we have dealt with the concept of an enzyme attached to a polymer matrix and the effects that the matrix imposes by virtue of its geometric and chemical structure on the function of the enzyme. It must be recognized, however, that while enzymes are attached to polymer matrices *in vivo*, these natural polymers can often undergo reversible changes in their structure. Thus membranes can fold/unfold, associate/dissociate and these processes may provide yet another level of control on the enzyme's activity. This enzymic regulation by alteration of the physical state of the polymer matrix has been ingeniously modelled by the attachment of enzymes to nylon fibres. Mechanical interference with the support can be transmitted to the attached enzyme by either a direct conformational change induced in the enzyme or by change in the microenvironment of the enzyme.

Trypsin has been attached with glutaraldehyde to nylon fibres previously subjected to partial acid hydrolysis. The enzymic activity of the trypsin falls dramatically when the fibre is stretched (see Figure 48) and returns to its original level when the fibre relaxes. This stretch/relax, activity decrease/increase cycle may be repeated many times. The fall in activity is unlikely to be due to a change in the microenvironment of the enzyme because variations in ionic strength (from 0.01 to 3) and pH (7–8.5) do not alter the magnitude of the mechanical effect. It should be noted also that if the substrate diffusion were limited, then stretching the fibre would 'expose' more enzyme and thus ought to raise the level of activity. One other factor is perhaps significant, the pattern of the fall of enzyme activity on stretching (Figure 48). Most of the loss of activity occurs when the fibre is stretched by just 1%. If we assume that the stretching of the fibre induces a similar conformational stretching in the protein molecule then, for trypsin with a diameter of approximately 50 Å, 1% stretching will cause a change of diameter of 0.5 Å, apparently sufficient to cause catalytic inactivation. The stretching of the fibre will of course only be transmitted to the protein molecule if there is more than one point of attachment of the protein to the fibre. Some of the trypsin molecules may be attached by only one glutaraldehyde bridge or two adjacent glutaraldehyde molecules; these would be unaffected

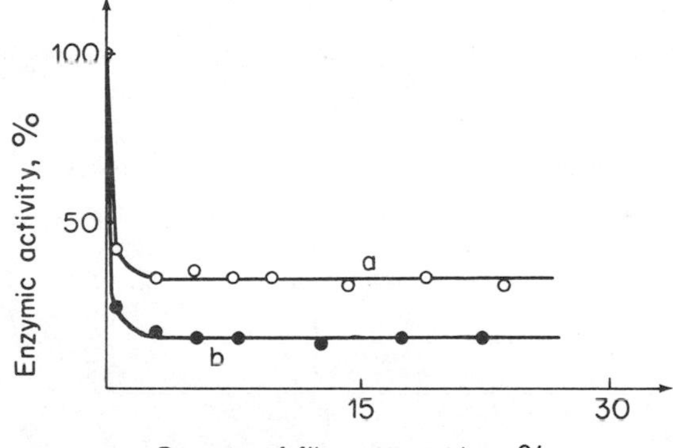

Figure 48. Dependence of enzyme activity of (a) α-chymotrypsin and (b) trypsin attached to nylon fibre, on the degree of stretching of the fibre. Substrates (a) *N*-acetyl-L-tyrosine ethyl ester (b) *N*-tosyl-L-arginine methyl ester. pH 8.0, 0.1 mol l^{-1} KCl at 25 ° C. Reproduced with permission from Berezin *et al.*, *Methods in Enzymology*, **XLIV**, 563 (1976)

by stretching of the fibre and would account for the constant residual activity shown between 2% and 30% stretching (see Figure 49). It is noteworthy that if the nylon–trypsin is reacted with *p*-nitrophenyltrimethyl acetate, a 'quasi substrate' that irreversibly acylates the enzyme (i.e. it does not undergo the final deacylation step in catalysis), then stretching of the fibre increases the rate of the reaction. This suggests that the stretch-induced conformational change occurring in the trypsin affects the catalytic groups involved in the deacylation step.

The stretching of the nylon fibres coated in trypsin can also show steric effects on enzyme–inhibitor interactions. Over a stretch range of 2–30% the trypsin does not show any change in activity in the absence of inhibitors but in the presence of high molecular weight inhibitors (e.g. pancreatic or soya bean trypsin inhibitors) the degree of inhibition increases sixfold as the fibre is stretched. With pancreatic trypsin inhibitor the effect is reversible on relaxation of the fibre. It can be deduced that this moderation of inhibition is due to the relaxed fibre forming a steric barrier between enzyme and inhibitor which is removed when the fibre is stretched (see Figure 49) exposing more catalytic sites. Oddly the effect of soya bean inhibitor (3.3 × MW of pancreatic inhibitor) is not reversed by relaxation. Presumably the larger soya bean inhibitor is trapped in the folds of the nylon fibre on relaxation and, by virtue of its size, is unable to escape into solution.

The moderation of enzyme activity by mechanical stress is similar to that observed by Ohnishi and Ohnishi (1963) who showed that when muscle fibres are stretched the inherent ATPase activity of the preparation increases several fold.

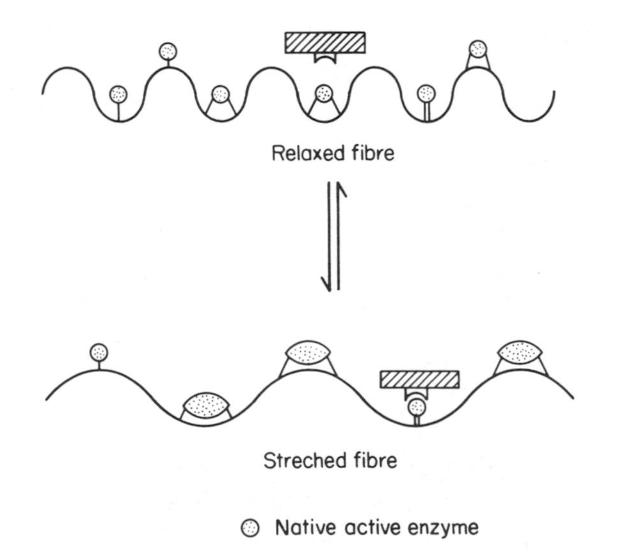

Figure 49. Schematic representation of the effect of stretching on the activity of nylon fibre immobilized enzyme

II MULTIENZYME SYSTEMS

The existence of naturally occurring organized multienzyme systems has been known for some time. Perhaps the best example of this is fatty acid synthetase, a complex of seven different protein molecules each repeated up to three times. These proteins have been identified as two distinct reductases, a hydratase, a malonyl transferase, a condensing enzyme, an acyl transferase, and a non-enzymic acyl binding protein. It can be deduced that the nature of the organization of these molecules allows a greater efficiency of fatty acid synthesis. This is not an isolated example. It now appears likely that almost any metabolic pathway may have its enzymes associated with some membranous structure of the cell. What evidence have we for this?

The glycolytic pathway is often assumed to exist as enzymes floating at random in the cytoplasm. However this is probably not the case, at least in certain tissues. Extraction of the enzymes of glycolysis from liver shows that when each enzyme is provided with the average cellular concentration of the relevant metabolites, the slowest enzyme reaction is the conversion of fructose-6-phosphate to fructose-1,6-diphosphate by phosphofructokinase (PFK). However the overall rate of the pathway *in vivo* is approximately 10% of the rate of the PFK step. A variety of hypotheses are possible.

1. The PFK is bound *in vivo* to a membranous structure, e.g. endoplasmic reticulum or the outside of the mitochondria (see p.125) which may be polyanionic. This will effectively lower the microenvironmental pH around the enzyme, both by a partitioning and diffusion effect (the reaction is acid liberating), thus lowering the activity of the PFK.

2. Such a polyanionic surface might restrict the assess of substrate to the enzyme, enhance the effect of an inhibitor, or alter the conformation of the enzyme molecule — PFK is an allosteric enzyme.

3. Such controls as in (1) or (2) above might even be exerted on other reactions in glycolysis.

4. The pathway is at some point interrupted by a piece of membrane, thus diffusion of one of the intermediates across the membrane becomes rate-limiting.

1. Order and Efficiency

It was in order to study such effects that Mattiason and Mosbach (1971) immobilized, at random, several enzymes capable of forming a mini pathway on to the same piece of cellulose. One such system consisted of the enzymes α-glucosidase, hexokinase, and glucose-6-phosphate dehydrogenase. In dilute solution these three enzymes will work in the presence of magnesium ions, ATP, maltose, and NADP to produce NADPH, but there is a considerable lag between addition of the substrates and the appearance of NADPH. When these three enzymes are bound to particulate Sepharose (by CNBr treatment) the lag between addition of substrates and NADPH production disappears (see Figure 50), despite the fact that the eventual rate of NADPH production is identical to the dilute solution experiment.

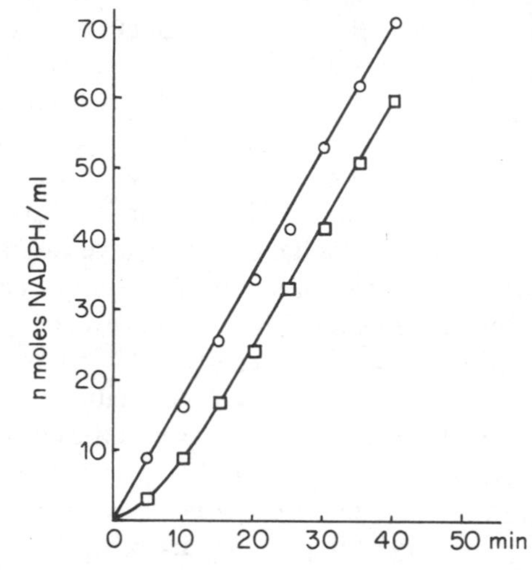

Figure 50. Illustration of time-dependent formation of NADPH by hexokinase/glucose-6-phosphate dehydrogenase system in dilute solution (□) and co-immobilized to cellulose (○). Reproduced with permission from Mattiason and Mosbach, *Biochim. Biophys. Acta*, **235**, 253 (1971)

It is assumed that the lag phase observed in the dilute solution experiment occurs because it takes a finite time for the concentration of the first product, glucose, to increase to a level where the hexokinase can operate and then longer still before sufficient glucose-6-phosphate is produced to allow operation of the glucose-6-phosphate dehydrogenase. The immobilized enzymes however have a built in advantage because the substrates for the second two enzymes (glucose and glucose-6-phosphate) are produced in the actual location of these enzymes, i.e. at the polymer surface, and therefore their effective concentration around the enzyme is always greater than the average concentration throughout the solution. Thus by localizing the enzymes in this way the pathway is made more efficient.

Obviously if the enzymes could be directly ordered (as opposed to randomly dispersed) on the polymer surface then the degree of efficiency can be effectively regulated.

Such localization of metabolic pathways will not just eliminate 'lag phases' it may also exert metabolic control, particularly if one of the components concentrated at the polymer surface is an enzyme activator or inhibitor.

Let us now return to the glycolytic pathway for a moment. The reactions in this pathway between PFK and pyruvate kinase are reversible. The enzymes that catalyse these reactions are often known as the 'constant proportion enzymes', because they are usually present in a given ratio. The reactions' equilibria are such that the tendency is towards the formation of fructose-1,6-diphosphate. This is a potent inhibitor of PFK. Clearly if pyruvate kinase activity falls then the level of fructose-1,6-diphosphate quickly builds up in the cell, particularly if the enzymes are located at a surface, and so PFK is rapidly inhibited, much faster and to a greater extent than it would be if the reactions occurred in dilute solution in the cytoplasm. Although this explanation is purely hypothetical, it might explain the 'slowed' glycolysis described above.

A number of other interesting immobilized enzyme systems have been described by Mosbach and co-workers and we shall consider just two of these. The first involves the co-immobilization of two enzymes which catalyse two consecutive reactions, the first of which is freely reversible. The two enzymes selected were malate dehydrogenase and citrate synthetase.

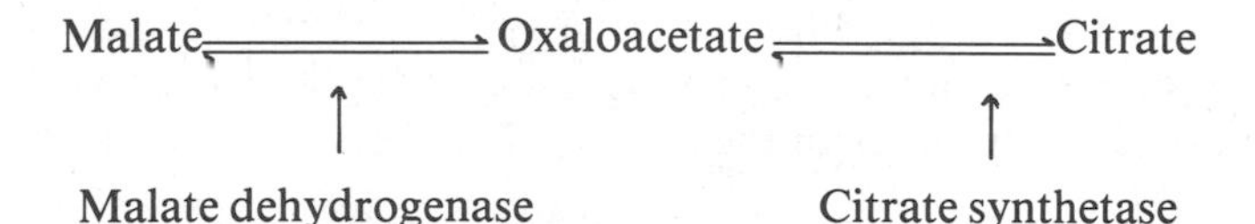

Left to its own devices, in dilute solution the first reaction favours the production of malate and thus in dilute solution the coupled enzyme reactions produce citrate at a slow steady-state rate. However, when they are co-immobilized the high local concentration of citrate synthetase effectively removes the oxaloacetate from the microenvironment of the malate dehydrogenase, preventing the reverse reaction, and hence enhancing the steady-state rate of production of citrate (see Figure 51).

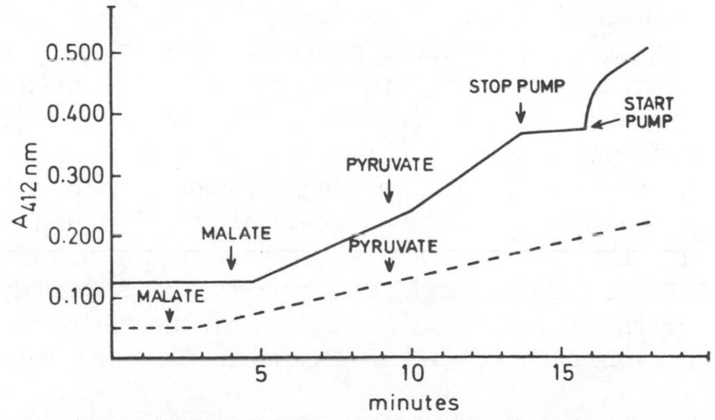

Figure 51. Time-dependent production of citrate by a three enzyme coupled reaction system, malate dehydrogenase–citrate synthetase–lactate dehydrogenase, co-immobilized in polyacrylamide (solid line) and in dilute solution (broken line). Arrows indicate additions. Addition of pyruvate elicits recycling of the NADH to NAD (by the lactate dehydrogenase) thus enhancing citrate production. Reproduced with permission from Srere *et al.*, *Proc. Nat. Acad. Sci. (US)*, **70**, 2534 (1973)

The second system reported from Mosbach's laboratories is more curious and involves the shift in pH optima observed for two consecutive immobilized enzymes. The enzymes used in this study were amyloglucosidase and glucose oxidase, the pH optima of which are 4.8 and 6.4 respectively.

When each enzyme is individually immobilized to Sepharose, neither pH optimum is altered. However when both enzymes are co-immobilized on to Sepharose the pH optimum of the combined reactions shifts to a higher value than that of the combined reactions operating in dilute solution. This is odd because it might be expected that in the immobilized system the microenvironmental glucose concentration will be greater and thus the glucose oxidase reaction faster than in the dilute solution system, which should result in the amyloglucosidase reaction becoming the rate-limiting step, causing a reduction in the observed pH optimum of the combined system.

2. Inhibitors as Activators

An even more intriguing multienzyme model system has been devised by Hervagault *et al.* (1975), in which an enzyme inhibitor apparently becomes an activator! This paradox is, however, very easy to explain and does raise considerable doubts about present ideas on feed forward inhibition and activation.

Hervagault's system used two consecutive reactions for the transformation of xanthine to allantoin via uric acid, reactions catalysed by xanthine oxidase and uricase respectively. Xanthine is usually a potent inhibitor of uricase activity. However, when a series of artificial membranes containing the same activity of uricase but varying xanthine oxidase activities are placed in solutions of differing xanthine concentrations, it can be shown that the degree of uricase activity (measured from allantoin production) is not only dependent upon the concentration of its inhibitor xanthine but also on the xanthine oxidase activity. The surprising result is that, provided the xanthine oxidase activity is high, increasing the xanthine concentration can apparently activate the uricase. When no xanthine oxidase is present, increasing the xanthine concentration, as one might expect, leads to a fall in uricase activity (see Figure 52).

Qualitative explanation is quite simple. With increasing levels of xanthine oxidase activity not only is the concentration of xanthine in the membrane reduced, but also the concentration of the product uric acid increases. Thus the concentration of the available substrate for the uricase increases above the control level and the uricase activity also rises above the control level.

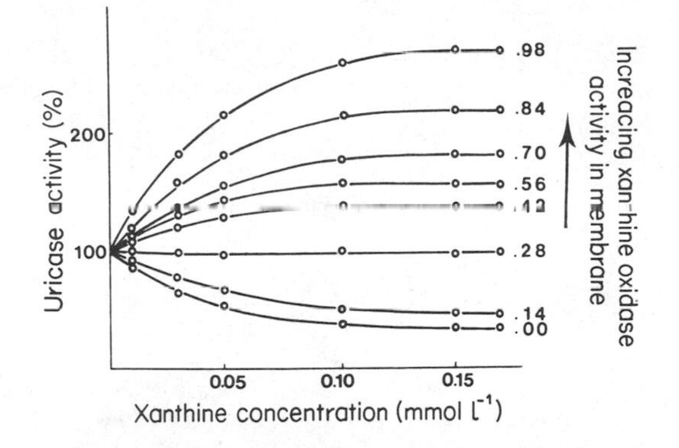

Figure 52. Relationship between xanthine concentration and uricase activity for uricase/xanthine oxidase membranes, each containing identical quantities of uricase activity but differing quantities of xanthine oxidase (0 to 0.98). Bulk phase uric acid concentration is the same for each system. Reproduced with permission from Hervagault *et al.*, *Eur. J. Biochem.*, **51**, 19 (1975)

Both these models illustrate one very important consideration. When the reactions of a cell are occurring in defined heterogeneous microenvironments, knowledge of the average concentrations of metabolites in the cell is insufficient even to begin to predict the pattern of inhibition, activation, and metabolic regulation in that cell.

III ACTIVE TRANSPORT

The realization of the technique of binding enzymes on to artificial membranes has given the scientist involved in the elucidation of the molecular process of active transport a valuable model of the biological situation.

1. Similarity is Justification

The similarity between proteins responsible for active transport across biological membranes and those responsible for enzymic catalysis has long been apparent. For example, both exhibit the characteristic of binding their 'substrate' to a specific site on the protein molecule and both exhibit similar kinetic characteristics. This similarity should hardly be surprising as both transport and enzymic proteins perform very similar functions, that is the 'conversion' of molecules; enzymes 'convert' molecules into different molecules, transport proteins 'convert' molecules from one location to another. The similarity does not end there. Certain classes of transport proteins also work as enzymes, chemically changing the molecules they carry across the membrane.

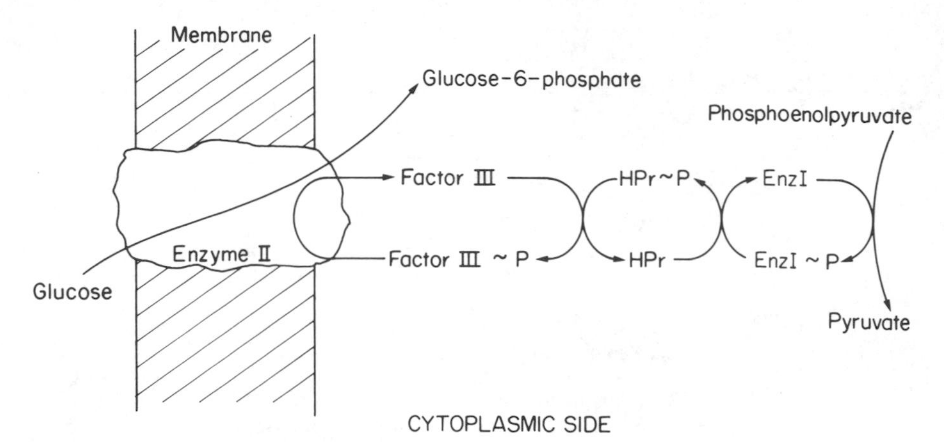

Figure 53. Group transfer of glucose by phosphoenolpyruvate-dependent translocating/phosphorylating system found in certain prokaryotes

An example of this is the passage of monosaccharides into certain bacteria. Here the molecule, for example glucose, is carried into the cell and simultaneously converted to glucose-6-phosphate as a necessary part of the process. This complex system is illustrated in Figure 53.

Alton Meister, in a review of the situation for the carriage of amino acids across the membranes of the small intestine, describes the process as an enzymic conversion (see Meister, 1973).

These observations are the justification for using enzyme membranes as model active transport systems.

Model systems in general, and models of active transport in particular, can be one of two types. The model may be very simple, thus its parameters may be easily defined. Such models, however, are usually structurally dissimilar to biological systems. Alternatively the model may be structurally similar to the biological system, in which case it is often as hard to define the parameters of the model as the biological system it mimics! Thus whatever the choice of model system great care must be exercised in its interpretation and application to the biological situation.

2. Simple Models

Most of the initiative in this area and other areas employing model systems came from the French workers Thomas and Broun. The earliest reported model of a facilitated transport system was for the transfer of bicarbonate through a thin silicone rubber membrane. Bicarbonate ions, in common with most other polar substances will not easily penetrate a silicone rubber membrane; carbon dioxide, however, will.

Thomas and Broun constructed an enzyme membrane from carbonic

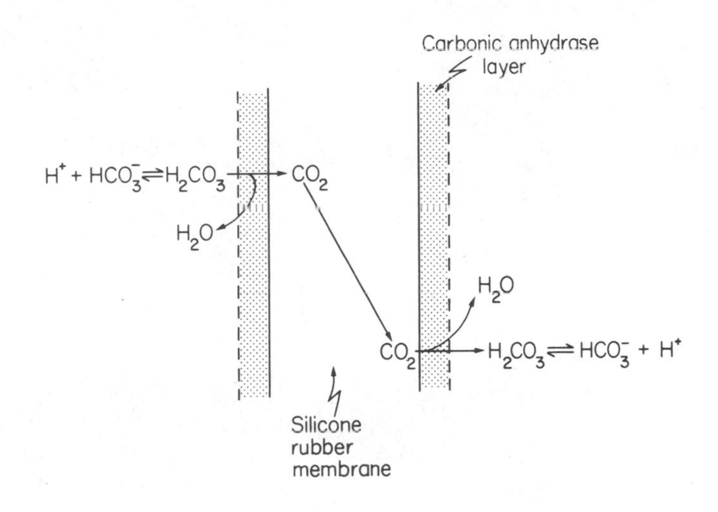

Figure 54. Silicone rubber/carbonic anhydrase membrane showing facilitated transport of bicarbonate

anhydrase and silicone rubber. The carbonic anhydrase was immobilized to both sides of the silicone rubber membrane by polymerization with glutaraldehyde (see Figure 54). When this was placed between two solutions, one of buffer and the other of buffer plus sodium bicarbonate, it was found that the presence of carbonic anhydrase markedly enhanced the rate of transfer of bicarbonate from one solution to the other. The bicarbonate ions are in equilibrium in solution with carbonic acid. As this molecule encounters the carbonic anhydrase on the surface of the membrane it is converted to carbon dioxide and water. The carbon dioxide moves freely through the membrane to the opposite surface where, because it again encounters a layer of carbonic anhydrase it is rapidly hydrated to carbonic acid thus to bicarbonate.

Active transport in the sense that the transported substance is accumulated unchanged against a concentration gradient has also been reported by Thomas and Broun. They constructed a complex membrane from alkaline phosphatase and hexokinase and used this to actively transport glucose. The membrane was composed of four layers, viz. a glucose-6-phosphate impermeable layer, a hexokinase layer, an alkaline phosphatase layer, and a final glucose-6-phosphate impermeable layer. The outer layers were in fact simple polyanionic membranes. The whole membrane was impregnated with ATP (Figure 55). When solutions of glucose of equal concentration are placed either side of this membrane the concentration of glucose on the alkaline phosphatase side is seen to increase at the expense of that on the hexokinase side (see Figure 56). This happens because of the glucose entering the hexokinase side of the membrane is immediately converted to glucose-6-phosphate. Glucose-6-phosphate cannot escape from the membrane and, because its concentration

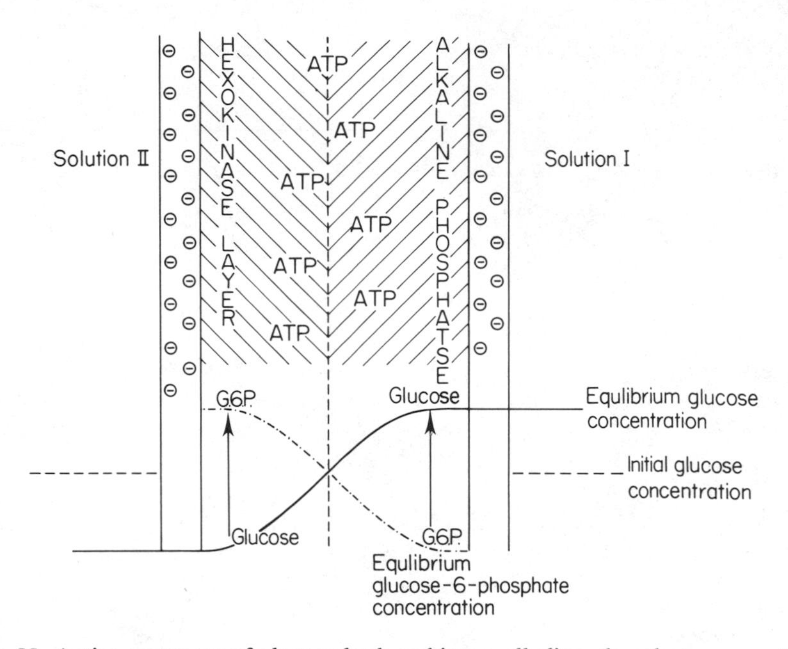

Figure 55. Active transport of glucose by hexokinase–alkaline phosphatase membrane impregnated with ATP. The outer polyanionic layers are impermeable to glucose-6-phosphate (see Figure 56)

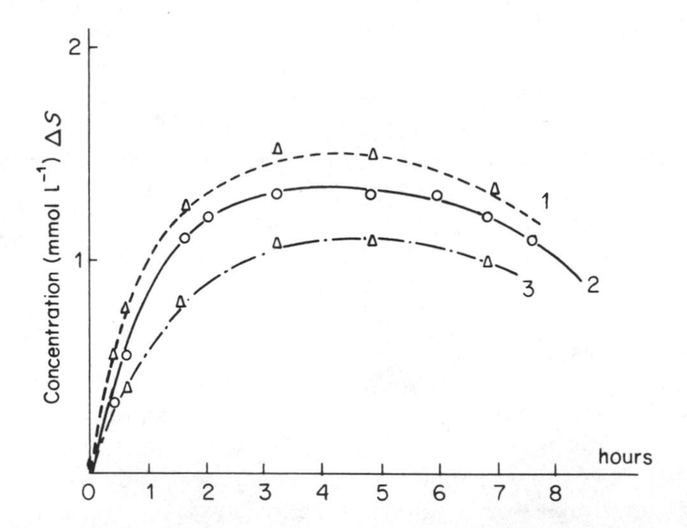

Figure 56. Time course of glucose accumulation by hexokinase–alkaline phosphatase membrane (Figure 55). ΔS is the difference in glucose concentration (solution I — solution II) for three values of initial glucose concentration (1) 1.12×10^{-2} mol l^{-1}, (2) 2.8×10^{-3} mol l^{-1}, (3) 5.6×10^{-3} mol l^{-1}. The decline of each curve after approximately 4 hours is a result of ATP depletion. Reproduced with permission from Broun *et al.*, *J. Membrane Biol.*, **8**, 313 (1972)

builds up around the hexokinase layer, it diffuses across the inside of the membrane to the alkaline phosphatase layer. There it is hydrolysed to glucose, which leaves by the shortest route, that is out of the alkaline phosphatase side of the membrane. As soon as the ATP is used up, the system returns to equilibrium. The obvious criticism of this system is that, in structure, it is wholly unbiological. It should be noted, however, that the asymmetry in the accumulation of glucose is a direct result not of expenditure of energy in the hydrolysis of ATP, but of the asymmetric nature of the membrane.

Another similar system, which is perhaps more comparable to the biological situation, uses an alkaline phosphatase/hexokinase membrane to actively accumulate glucose-6-phosphate. A polymethacrylic acid membrane containing homogeneously distributed alkaline phosphatase and hexokinase separates two equimolar solutions of glucose-6-phosphate; one solution also contains ATP (see Figure 57).

The glucose-6-phosphate concentration on the ATP side of the membrane rises at the expense of that on the other side (Figure 58). On the non-ATP side of the membrane glucose-6-phosphate, which cannot penetrate the membrane very easily whereas glucose can, is converted to glucose, the concentration of which rises at that edge of the membrane. On the ATP side of the membrane any glucose-6-phosphate converted to glucose is immediately reconverted to glucose-6-phosphate by the hexokinase. Thus glucose tends to diffuse across the membrane, towards the ATP side, down a concentration gradient. The glucose concentration is at all times significantly lower than the glucose-6-phosphate concentration and hence is not observed as an intermediate.

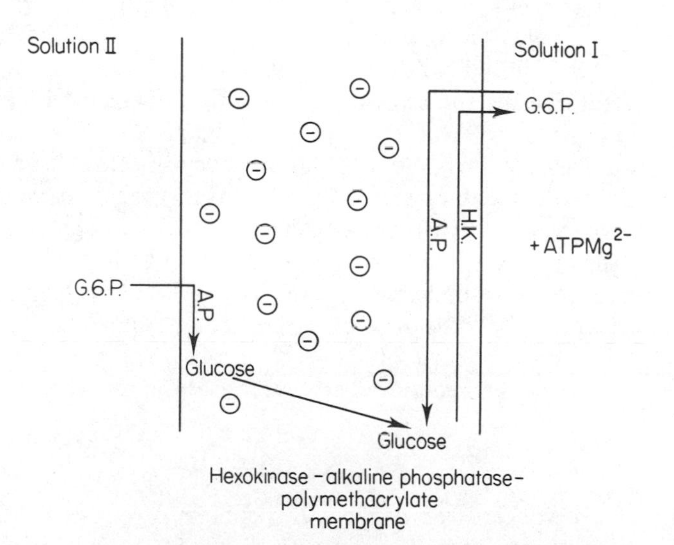

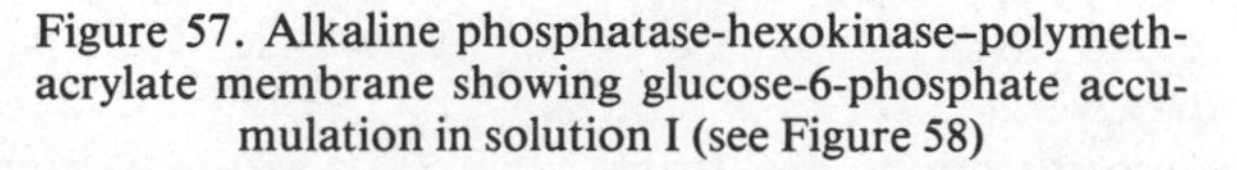
Figure 57. Alkaline phosphatase-hexokinase-polymethacrylate membrane showing glucose-6-phosphate accumulation in solution I (see Figure 58)

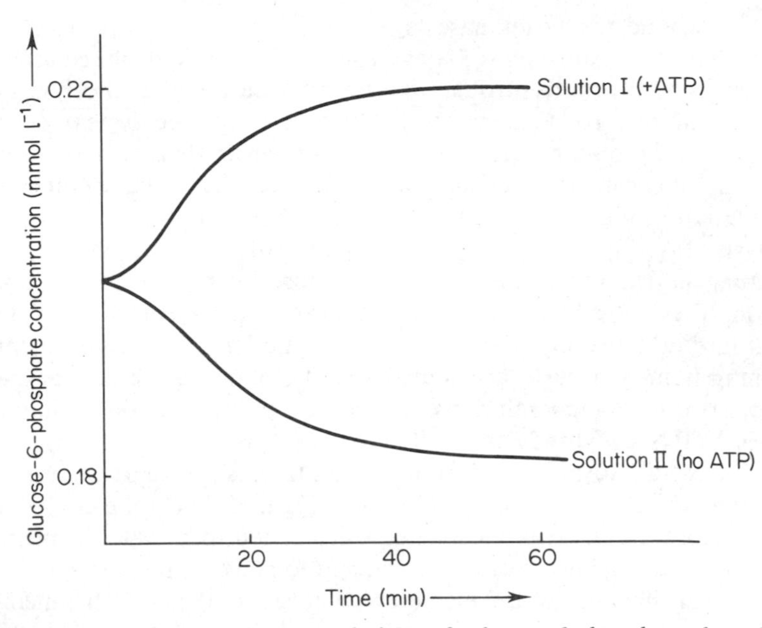

Figure 58. Time course of accumulation of glucose-6-phosphate by alkaline phosphatase–hexokinase–polymethacrylate membrane (see Figure 57)

In this case the apparent active transport depends for its effect not on metabolic energy (i.e. ATP) nor on the asymmetry of the membrane, but on the existing asymmetric distribution of the metabolite ATP across the membrane. ATP is required for this transport process, not to act as an energy source *per se*, but to keep the reaction cycle going. Transport takes place by simple diffusion.

The two examples explained above both show one important aspect of active transport, that is the accumulation of a substance against a concentration gradient with concomitant hydrolysis of ATP. They also demonstrate that the asymmetry of the system, rather than the release of chemical energy from the hydrolysis of ATP, is responsible for this accumulation. Neither system, however, can transport a substance at a rate much faster than it is able to diffuse through the membrane of its own accord. For a system that can accumulate substances at an enhanced rate we must look elsewhere.

3. Complex Models

Storelli and his co-workers approached the problem of constructing a model active transport system from the opposite direction, that is they chose to construct a biologically accurate model utilizing biologically derived materials. The enzyme they selected for their studies was an isomaltase–sucrase enzyme known to be associated with and extracted from the epithelial membrane of the duodenum. They incorporated this enzyme into a mixture of purified phospholipids and made an artificial lipid membrane (a black lipid membrane) containing the enzyme, by painting their lipid-enzyme solution over a small hole in a teflon plate which separated two compartments. The lipid mixture thins down spontaneously to form a single lipid bilayer under these conditions (Figure 59) within which the enzyme is incorporated. Sucrose added to one side of the membrane rapidly disappeared and fructose appeared at the same rate on the other side, demonstrating the simultaneous hydrolysis and transport of sucrose.

The rate of transaccumulation of the fructose was several orders of magnitude faster than predicted by a consideration of the diffusion coefficients of fructose, glucose, and sucrose through a lipid bilayer, either in the absence of isomaltase–sucrase or in the presence of inhibited isomaltase–sucrase; to all intents and purposes the membrane in these states was impermeable to all three sugars. Several other facets of the system were also observed. The sucrose, when added to one side of the active isomaltase–sucrase membrane did not diffuse through to the other side, nor would the fructose diffuse back. Membranes prepared from pure phospholipid to which the isomaltase–sucrase is subsequently attached by adding the enzyme to the solution on one side of the membrane, displayed no translocating ability, even though the enzyme was still active. Thus it was possible to conclude that for active translocation of the sucrose (with concurrent hydrolysis) the enzyme must be active and penetrate the membrane.

This model demonstrated that an enzyme when arranged so that it penetrates the membrane can show translocating ability, the transport depending upon, in this case, the hydrolysis of the sucrose. Exactly how this

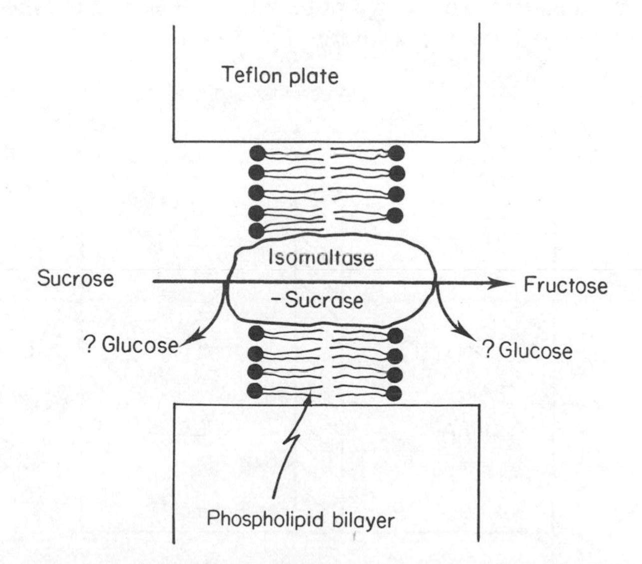

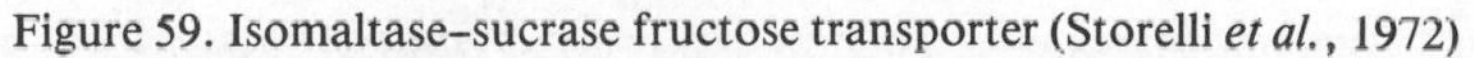
Figure 59. Isomaltase–sucrase fructose transporter (Storelli *et al.*, 1972)

model works is still something of a mystery. The mechanism of action of other sucrases has been shown to be a three-step process, binding of sucrose, hydrolysis and release of fructose, and finally release of glucose from the enzyme. Presumably during the catalytic process once the sucrose has been hydrolysed, a conformational change must take place in the enzyme, causing the active site to move and become open to the other side of the membrane. The thermodynamic requirement for energy input to balance the potential energy gained by the asymmetric accumulation of fructose is provided by the decrease in potential energy due to the diminution of the sucrose gradient or perhaps by the release of energy from the hydrolytic reaction.

IV HETEROGENESIS

We can now turn to a consideration of other model systems which either generate or utilize heterogeneity. Within this section we will consider systems generating electrical current, demonstrating hysteresis, oscillatory phenomena, and the spontaneous generation of heterogeneity.

1. Current Generation

David *et al.* (1974) has described a system comprising a symmetrical urease polyionic membrane capable of generating an electrical potential. The urease membrane was placed between equimolar solutions of urea, both solutions were buffered to pH 5.1, one with 10^{-2} mol l^{-1} phosphate the other with 10^{-3} mol l^{-1} phosphate (Figure 60). Thus the essential asymmetry of the system is contained in the different buffering capacities of the two solutions. It was found that such a system generated a potential of 40 mV across the membrane which was eliminated on the addition of the urease inhibitor, bisulphite

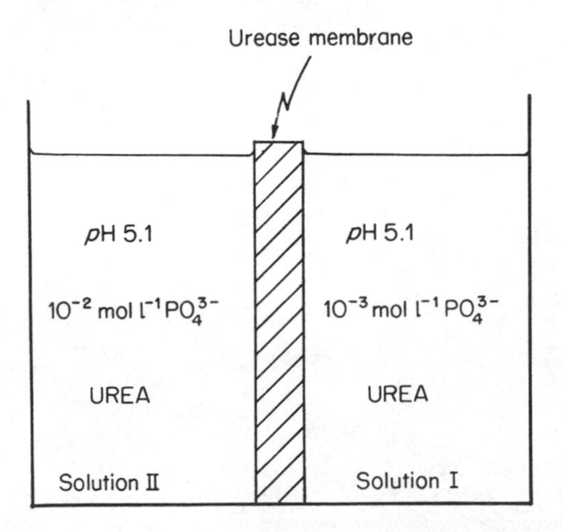

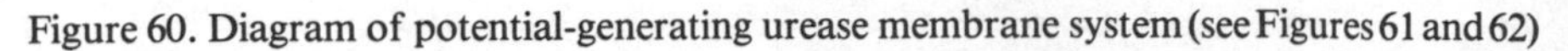
Figure 60. Diagram of potential-generating urease membrane system (see Figures 61 and 62)

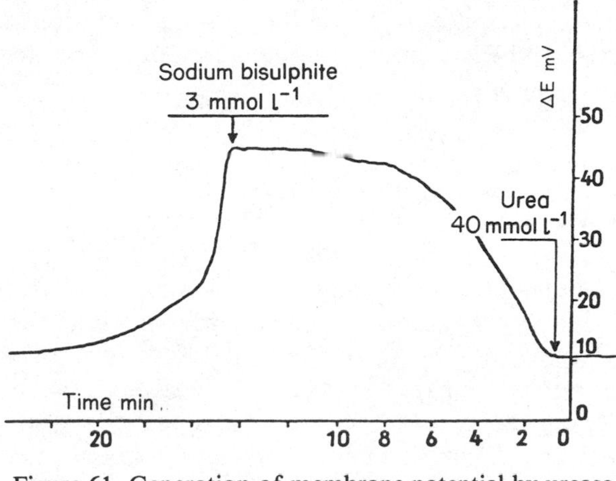

Figure 61. Generation of membrane potential by urease membrane as a function of time in the presence of urea and sodium bisulphite (urease inhibitor) (see Figure 60). Reproduced with permission from David *et al.*, *J. Membrane Biol.*, **18**, 113 (1974)

(Figure 61). The value of the potential developed depended upon the urea concentration (Figure 62) this dependence being sigmoidal, a characteristic associated with cooperativity in enzymes. A qualitative explanation can be proposed. The reaction initially proceeds equally but, as at pH 5.1 it is alkali liberating, the pH in the solution with the lower buffering capacity (solution I) rises more rapidly than that with the higher buffering capacity (solution II). The membrane then begins to act asymmetrically as the reaction on side I will

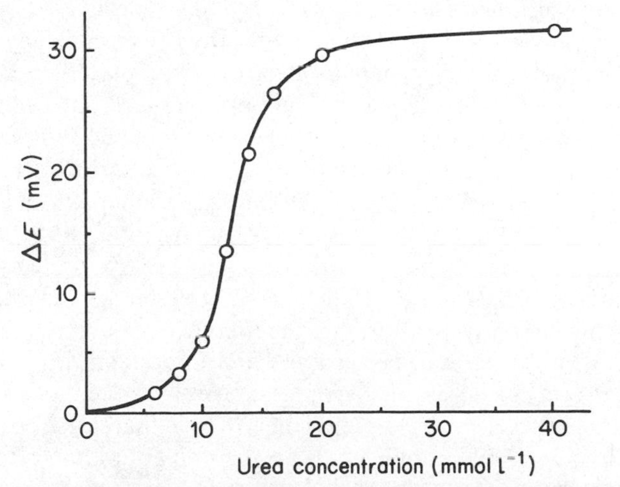

Figure 62. Steady-state membrane potential generated by urease membrane as a function of urea concentration (see Figures 60 and 61). Reproduced with permission from David *et al.*, *J. Membrane Biol.*, **18**, 113 (1974)

proceed at a faster rate than that on side II, thus solution II will come to contain a greater concentration of products, ammonium ions and carbonic acid, than solution I, the ionic imbalance causing a potential difference across the ion-selective membrane.

2. Hysteresis and Memory

The ability to remember depends upon the existing state being a result not only of present environmental conditions but the way in which these have changed. In the context of enzymology the activity of an enzyme will depend not only upon a particular environmental factor but whether that factor has been increasing or decreasing. The environmental factor may be anything that affects the rate of an enzyme reaction in a non-linear fashion, for example pH or substrate concentration where substrate inhibition is also present. The existence of hysteresis phenomena in immobilized enzymes subject to substrate inhibition has been discussed (Chapter 2, Section IV, 4, d). Naparstek has described a system where the microenvironmental pH within an immobilized papain preparation at a given bulk phase pH, can have one of two values, depending upon whether the bulk phase pH is being raised or lowered. Such responses are known as hysteresis effects and are commonly observed in the field of engineering. It has for a long time been assumed that hysteresis phenomena must occur in biological systems, the common proposition being that they were the result of structural changes within macromolecules. The beauty of Naparstek's system is that no structural changes take place, the hysteresis observed resulting from the diffusional constraints placed upon hydrogen ions by the polymer matrix. The system consisted of a combined pH-electrode coated with a thin layer of cellulose nitrate. Papain was cross-linked into the cellulose nitrate with glutaraldehyde. The pH registered by this papain electrode was a measure of the pH of the microenvironment of the immobilized papain pH_i. The papain electrode was placed in a solution of substrate benzoyl arginine ethyl ester, the bulk phase pH of which, pH_e, was measured with another pH electrode (Figure 63). Figure 64 shows the effect on pH_i of raising and lowering the value of pH_e by the addition of alkali or acid. The hydrolysis of benzoyl arginine ethyl ester by papain has a pH optimum in free solution of about 8.0 and releases benzoyl arginine, ethanol, and a H^+ ion, thus, because of the diffusion limitation imposed upon the hydrogen ions by the polymer matrix, the pH_i of the reacting system is always lower than pH_e. At a sufficiently high pH_e, despite the presence of substrate, the values of pH_i and pH_e will be identical, because the enzymic rate at this pH is zero. If pH_e is slowly lowered the enzyme becomes active, hydrogen ions are produced in the polymer matrix decreasing pH_i, which in turn increases the reaction rate and thus the value of pH_i rapidly falls to the pH optimum for papain. Any further reduction in pH_i becomes self-limiting because it would result in a decrease in enzymic rate. However, if the reaction is initiated at a pH_e of 8 (the pH optimum for papain) the value of pH_i will also be 8. If pH_e is now raised

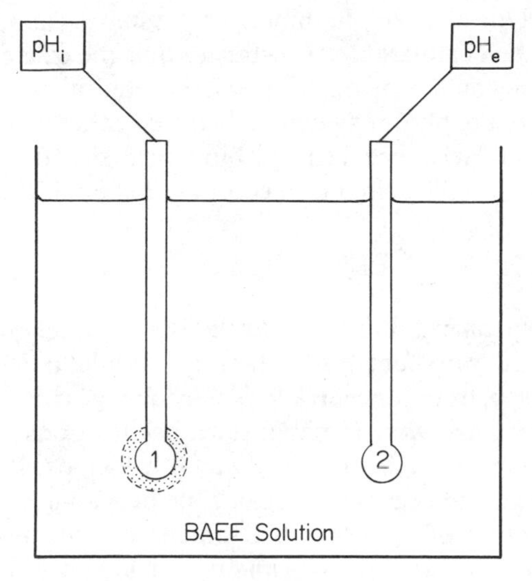

Figure 63. Measurement of microenvironmental pH (pH_i) of immobilized papain. (1) papain–pH electrode. (2) pH electrode. (see Figure 64)

slowly the value of pH_i will remain at 8 until a sufficiently large H^+ ion concentration gradient is produced, such that the potential rate of outward diffusion of the hydrogen ions can exceed their rate of production. At this point the reaction rate will fall rapidly and the value of pH_i will suddenly increase. The resulting pattern will be observed as a hysteresis curve.

The potential importance of the hysteretic behaviour of microenvironmental

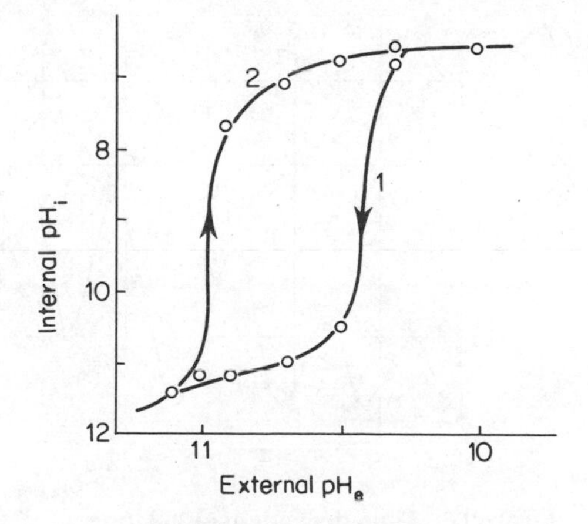

Figure 64. Hysteresis relationship between pH_i and pH_e for immobilized papain–pH electrode (Figure 63) as a function of (1) increasing and (2) decreasing pH_e. Reproduced with permission from Naparstek *et al.*, *Nature*, **249**, 490 (1974)

pH as a control mechanism in biological systems cannot be stressed too strongly. It may be postulated, for instance, that the existence of bistable pH states within a cell would provide not just a mechanism for switching on or off one particular enzyme, but for switching between entire metabolic pathways or sequences. The similarity between the biological situation and the mode of action of bistable switching circuits in computers should be apparent.

3. Oscillations

Hysteresis phenomena can thus be produced in biological systems as a result of changes in certain rate-controlling factors. While it is easy to visualize situations where substrate concentrations may change over a period of time (in fact oscillations in *in vivo* metabolite concentrations have been clearly demonstrated in glycolysis) is there any way in which oscillations in pH might be produced in reacting enzyme systems, thus providing a mechanism for the generation of the type of hysteresis phenomena described above? The answer to this question lies in another experiment by Naparstek and his co-workers. They preincubated a papain–pH electrode in buffer at pH 6 until the pH_i registered that value, then transferred it to a solution of benzoyl arginine ethyl ester buffered at pH 9.3. They found that the value of pH_i gradually increased from 6 to 8 in an oscillatory fashion (Figure 65), the oscillations having a period of about 30 seconds. This uncharacteristic behaviour of an otherwise simple enzyme is once again brought about by the restriction of the free diffusion of hydrogen ions by the polymer matrix to which the papain is immobilized. When the papain electrode is subjected to a pH_e of 9.3, hydroxyl ions diffuse into the immobilized enzyme's matrix thus raising the value of pH_i.

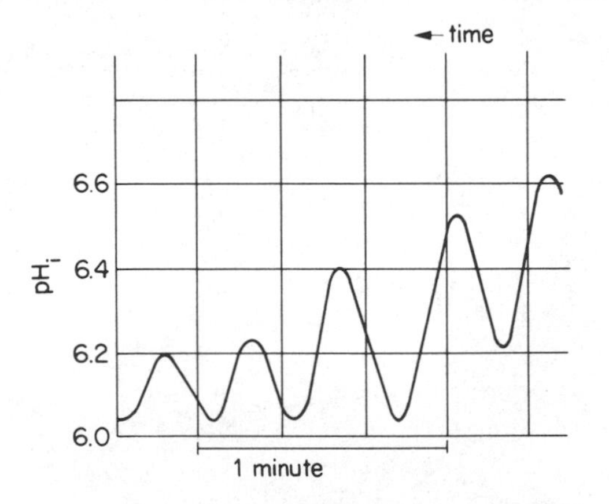

Figure 65. Time-dependent oscillation of pH_i of papain–pH electrode immersed in substrate solution (benzoyl arginine ethyl ester) at pH_e 9.3. Reproduced with permission from Naparstek *et al.*, *Biochim. Biophys. Acta*, **323**, 643 (1973)

This in turn leads to an increase in the rate of the reaction and hydrogen ion generation the latter causing pH_i to fall. The fall in pH_i increases the hydroxyl ion concentration gradient and decreases the reaction rate, such that the rate of diffusion of hydroxyl ions into the polymer matrix exceeds the rate of reaction. Thus the pH_i again rises causing the reaction rate to increase. This cycle of events recurs repeatedly, the minimum value of pH_i increasing slightly with each oscillation.

The type of oscillatory phenomenon observed depends upon a sudden change in the bulk solution pH. If this type of oscillatory phenomenon is to be considered as a serious model of *in vivo* behaviour, the spontaneous enzymic generation of heterogeneity of pH must also be demonstrable. For an example of this we must turn to the work of Thomas and his co-workers. They constructed a symmetrical membrane containing a homogeneous mixture of glucose oxidase (pH optimum 5) and urease (pH optimum 7). Below a pH of 9.0 the glucose oxidase catalysed reaction will release hydrogen ions, the urease catalysed reaction releases hydroxyl ions. The membrane separates two identical solutions of a pH between 5 and 7 containing glucose and urea, whose concentrations relative to each other are adjusted such that, on reaction, the total acid produced by the membrane-bound glucose oxidase reaction equals the total alkali produced by the urease reaction. At time zero a small pH gradient ($\pm$ 0.05) is created across the membrane and if no further constraints are applied to the system, a pH gradient of up to 2 units will rapidly develop across the membrane. The exact value of this gradient will depend upon a number of factors, for example initial pH_e, relative and absolute maximal reactivity of the enzymes, substrate concentrations, and the permeability of the membrane to H^+ ions. The initial, small pH gradient created will result in an imbalance of reaction rate of the two enzymes such that, on the side of the membrane where the pH_e is raised, the activity of the urease will be increased (and that of the glucose oxidase decreased). Thus the rate of hydroxyl ion production is increased and exceeds the rate of proton production by the glucose oxidase reaction. This causes the pH_e to rise yet more, increasing the imbalance between the rate of hydroxyl and hydrogen ion production. Thus the increase in pH is autocatalytic up to a maximum of the pH optimum for the urease reaction. Conversely on the other side of the membrane the pH_e will fall to the pH optimum for the glucose oxidase reaction by a similar autocatalytic mechanism.

Thus it is theoretically possible to propose that rapid heterogeneous distribution of hydrogen ions within a biological system may give rise to local oscillations in pH, which in turn may result in hysteretic changes in the pH of a particular microenvironment. Such hysteresis phenomena could perhaps play a sophisticated role in the control of cell metabolism and hence cellular function.

V BIOLOGICAL RELEVANCE OF MODEL IMMOBILIZED ENZYME SYSTEMS

We have seen in the preceding sections of this chapter and also in Chapter 2, that enzymes acting in a heterogeneous environment take on characteristics and attributes, unavailable to the same enzymes acting in dilute solution, that could play an important role in the control of cellular metabolism and cellular function. It is pertinent therefore to ask how relevant the knowledge gained of this peculiar behaviour of enzymes is to an understanding of the way in which the cell operates.

Certain similarities are obvious, most membranes are polyanionic, many enzymes are bound to the membranes of the cell with varying degrees of affinity or are associated into large multienzyme complexes. Some enzymes, however, behave in the cell as though they are in dilute solution. For example, xanthine oxidase exhibits the same activation energy and K_m both *in vivo* in lymphocytes and *in vitro* in dilute solution. One is forced to conclude therefore, that the environment of the lymphocyte's cytoplasm is not subjected to diffusional limitations, despite the presence of large quantities of endoplasmic reticulum. This is perhaps surprising and one of three further conclusions must be drawn. (1) For some reason the extrapolation of concepts derived from artificially immobilized enzymes to the *in vivo* state do not apply in this case. (2) The endoplasmic reticulum does not present a diffusional barrier to xanthine. (3) The endoplasmic reticulum does not exist and is an artefact of electron microscopy. Although this latter conclusion is controversial to say the least, it has been proposed by Hillman and Sartori and is based on geometric considerations and observations of the living cell. It is mentioned here, not because the author necessarily supports this view, but because, if true, it would be the most likely explanation of the *in vivo* behaviour of xanthine oxidase.

More useful comparisons can be drawn between the artificial immobilized enzyme systems containing more than one coupled enzyme reaction (Chapter 4, Section II) and naturally occurring multienzyme systems. However, direct comparisons are limited because co-immobilized enzymes are attached in a random fashion, whereas in biological multienzyme complexes the enzymes are usually in a precisely ordered sequence, with consequent enhancement of the efficiency of the system. This is particularly so in large cells which may require some mechanism to ensure sufficient enzyme substrate collisions.

Perhaps the most useful lesson to be learnt from the artificial system relates to the nature of the pH dependence displayed by multienzyme complexes and the manner in which it may be modified by the alteration of the catalytic activity of individual enzymes (Chapter 4, Section II). It is much more difficult to compare the perturbation of an enzyme's pH dependence brought about by immobilization and effects that may be present *in vivo*. One reason for this is that pH is a statistical concept and has no relevance in microscopic particles, whose volume may be so small that each particle contains only a few hydrogen ions. For instance a particle with a diameter of 0.5 μm at a nominal pH of 7.0 will contain only 4 hydrogen ions. In such circumstances the localization of charge within a particle will be of paramount importance. If an enzyme contains a catalytically active carboxylate ion then any change in its proximity to a protonated amine group will change the local electrostatic charge around it, affecting the pK_a (of both groups) and thus the enzyme's pH dependence.

This in effect is what happens when the surface charge of an enzyme is modified (see Chapter 2, Section III. 7). Thus observation of an enzyme attached to a polyanionic surface *in vitro* may produce results in direct contradiction to those expected from the study of the enzyme attached to a polyanion *in vivo*. One example of this is the enzyme aconitase which, in dilute solution *in vitro* displays a pH optimum above 7. *In vivo* aconitase is attached to the mitochondrial membrane which in common with most biological membranes is polyanionic. It might be expected therefore that *in vivo* aconitase will display a higher pH optimum than it does *in vitro* because of accumulation of hydrogen ions at the membrane surface, however in practice exactly the reverse is true; mitochondrial-bound aconitase has a pH optimum of less than 6. The explanation of this observation must be a matter of pure speculation. For example the aconitase may not be located at the surface of the membrane but may be partially buried; the enzyme may be specifically located in a region of the membrane exhibiting a net positive charge; substrate diffusion to the enzyme may be pH dependent and this may be superimposed upon the true pH optimum of the aconitase. Which, if any (or all), of these is the genuine explanation it is impossible to ascertain. We are led to the inescapable conclusion that *in vivo* pH dependence of enzymes may be far more complex than even immobilized enzyme studies suggest. This has considerable implications when attempting to estimate the catalytic activity of enzymes acting *in vivo*. Obviously determination of the cellular activity of an enzyme by assaying it at the 'average' cellular pH, usually taken to be 7.4, is liable to be highly erroneous. Thus great caution must be exercised in assigning *in vivo* metabolic function to an enzyme on the basis of its acivity determined by such methods. The example given in Chapter 4, Section II of the overall rate of glycolysis may be the result of just such an error. An enzyme's activity *in vivo* will be affected not only by local variations in pH or electrostatic charge, but may also be affected by changes in distribution of the enzyme. Hexokinase exists in the cell both in the cytoplasm as the 'free' enzyme and bound to the mitochondrion, an equilibrium existing between these two distributions. When it becomes bound to the mitochondrion the hexokinase exhibits a fivefold increase in catalytic activity. This cannot be due to the presence of a negatively charged surface altering pH dependence or rejecting the product inhibitor glucose-6-phosphate, because hexokinase adsorbed on to small negatively charged paraffin spheres shows a decrease in catalytic activity. It may be that the binding of the hexokinase to the paraffin spheres causes a disruption in the tertiary structure of the hexokinase, or the non-specific nature of the association results in the hexokinase being bound upside down. It can be demonstrated that treatment of hexokinase with α-chymotrypsin results in the excision of a small peptide and a loss of the ability of hexokinase to become bound to mitochondria. This would suggest that the binding process involves a specific interaction between enzyme and organelle. The possibility that the mitochondrion may provide hexokinase with a high local concentration of ATP can also be ruled out, because all the ATP available to the bound hexokinase is equally available to glycerokinase acting in the solution in which the mitochondria are suspended. Perhaps the enhancement of activity is brought about by the removal of ADP into the mitochondrion or, alternatively, it may be that the binding process induces a beneficial conformational change in the hexokinase. What is clear is that the metabolic activity of the hexokinase may be effectively controlled by altering its distribution between the soluble and bound forms.

Localized conditions within the cell may affect the nature of an enzyme reaction. For example glucose-6-phosphatase is buried in the endoplasmic reticulum which imposes considerable restriction on the free diffusion of the substrate to the enzyme. Furthermore, glucose-6-phosphatase can catalyse the phosphorylation of glucose in the presence of high concentrations of an effective phosphate donor such as carbamoyl phosphate or phosphoenolpyruvate. Thus the possibility exists that at one face of the membrane glucose-6-phosphatase may act as a phosphatase while at the other face, given appropriate high local concentrations of a suitable phosphate donor, it may act as a phosphorylase. Whether it could act as a membrane transporter for glucose-6-phosphate in a fashion similar to that described in Section III of this chapter is perhaps more questionable.

While great caution must be exercised in extrapolating from model immobilized enzyme systems to the *in vivo* state, certain conclusions may be drawn. The metabolizing cell is more than just a bag of soluble enzymes. The enzymes have the ability to create a heterogeneous environment within the cell and this heterogeneity may in turn be used to control the metabolic processes occurring within the cell. Thus we are left with the concept of the control of cellular metabolism by a system, the complexity of which is several orders of magnitude greater than our existing concepts, and which depends upon the imposition of spatial on the known temporal effects. The matter in question may be summarized in the words of McLaren and Packer who wrote 'As soon as we encounter enzyme action in films, mitochondria, cell surfaces, and membranes we encounter in a dramatic way the question of the meaning of concentration in volume elements so small as to raise doubt about the adequacy of statistical averages of numbers of molecules and ions. This in turn raises doubts about the validity of mass action, equilibrium, steady-state assumptions and so forth, when ions, particularly hydrogen ions are being formed or consumed or local charges are altered in position and number by virtue of reorganization of subcellular structures.' Equally we could end this discussion by reflecting on the conclusion of Munkres and Woodward; 'Evolution of intracellular localization of enzymes may have led to a more efficient metabolic system and may also have allowed a new enzyme function: vectorial catalysis.'

Chapter 5

Immobilization Methods

This chapter contains six methods for enzyme immobilization, adapted from various sources. The references quoted at the end of each method are the principal sources of information. When selecting a method for immobilization of an enzyme the proposed use of the preparation should always be born in mind (see Chapter 1, Section VI).

I ELECTROSTATIC BONDING OF AMYLOGLUCOSIDASE TO DEAE-SEPHADEX

In principle the simplest form of enzyme immobilization is that involving electrostatic bonding of enzyme to polymer. This method is generally applicable so long as the enzyme and polymer matrix bear opposite charges. The greatest stability of bonding occurs when the isoionic point of the enzyme is furthest removed from its operational pH range. Care must be taken to control carefully the pH and ionic strength to which the immobilized enzyme preparation is subjected, in order to prevent desorption of the enzyme.

Materials

Amyloglucosidase solution
(5 g in 100 cm^3 of 0.02 mol l^{-1} acetate buffer pH 4.2)
DEAE-Sephadex Type A 25 (10 g)
Acetate buffer 0.02 mol l^{-1} pH 4.2
HCl 0.5 mol l^{-1}
NaOH 0.5 mol l^{-1}

Method

1–4 Preparation of DEAE-Sephadex
5–8 Coupling of enzyme

1. Suspend 10 g DEAE-Sephadex in 50 cm^3 0.5 mol l^{-1} HCl and stir for 20 min.
2. Filter, resuspend in 50 cm^3 of 0.5 mol l^{-1} NaOH and stir for 20 min.
3. Filter, wash with distilled water until washings are neutral, suspend in

100 cm^3 acetate buffer pH 4.2 and stir for 10 min.

4. Filter, resuspend in 100 cm^3 acetate buffer and leave to equilibrate overnight. Filter dry.
5. Add the prepared DEAE–Sephadex to 100 cm^3 of the amyloglucosidase solution, stir at room temperature for 2 hours and then filter.
6. Reslurry the enzyme–Sephadex complex in 200 cm^3 of acetate buffer pH 4.2, stir for 15 min and filter.
7. Repeat step 6 until the filtrate is free of protein as judged by its absorbance at 280 nm.
8. Resuspend the amyloglucosidase-Sephadex complex in 50 cm^3 of acetate buffer and store at 4 °C.

Tosa, T., Mori, T., Fuse, N., and Chibata, I. (1967) *Enzymologica*, **32**, 163.

II IMMOBILIZATION OF α-CHYMOTRYPSIN TO CYANOGEN BROMIDE ACTIVATED SEPHAROSE

Cellulose, agar, or agarose may all be reacted with cyanogen bromide to form an activated polymer capable of bonding protein. After reaction with protein the remaining imido-carbonate residues are deactivated by reaction with glycine. The pH of the cyanogen bromide–Sepharose reaction mixture must be kept alkaline in order to avoid the evolution of cyanide. The pH may be controlled, either by the use of buffers or by the monitored addition of alkali.

Materials

Sepharose 4B
Phosphate buffer 2.0 mol l^{-1} pH 12.1
Phosphate buffer 5.0 mol l^{-1} pH 12.1
Cyanogen bromide solution
(100 mg in 1 cm^3 phosphate buffer 5.0 mol l^{-1} pH 12.1)
α-Chymotrypsin (100 mg dissolved in 4 cm^3 of 0.3 mol l^{-1} $KHCO_3$)
Glycine
$NaHCO_3$ solution 0.1 mol l^{-1}
HCl 10^{-3} mol l^{-1}
NaCl 0.5 mol l^{-1}

Method

1–5 Preparation of activated Sepharose 4B
6–9 Coupling of α-chymotrypsin to activated Sepharose

1. Wash 10 g Sepharose 4B with 2.0 mol l^{-1} phosphate buffer pH 12.1 and filter dry on a Buchner funnel.
2. Slurry the Sepharose in 10 cm^3 cold phosphate buffer 5.0 mol l^{-1} pH 12.1 and add 20 cm^3 of cold distilled water.
3. To the Sepharose slurry, at 5 °C, add the cyanogen bromide solution

dropwise, such that the whole volume is added within 5 min.
4. Stir the mixture for a further 10 min.
5. Transfer the gel to a glass filter and wash with 100 cm^3 cold distilled water.
6. Suspend the activated Sepharose in 50 cm^3 $NaHCO_3$ and add to the solution of α-chymotrypsin.
7. Stir gently for 24 hours at 4 °C.
8. Add 1.5 g glycine to the slurry and stir for a further 24 hours.
9. Filter the α-chymotrypsin–Sepharose complex and wash in turn with 200 cm^3 aliquots of water, 0.1 mol^{-1} $NaHCO_3$, 10^{-3} mol l^{-1} HCl, 0.5 mol l^{-1} NaCl, and finally distilled water.

Porath, J., Aspberg, K., Drevin, H., and Axen, R. (1973) *J. Chromatogr.*, **86**, 53.

III IMMOBILIZATION OF PAPAIN TO CYANURIC CHLORIDE ACTIVATED CELLULOSE

Cyanuric chloride (trichloro-s-triazine) will react rapidly with free hydroxyl groups present on cellulose. The dichloro-*s*-triazyl-cellulose may then be reacted with the enzyme and, finally, the third chloro group substituted with an amine or other chemical group. However, the dichloro-*s*-triazyl compound is fairly reactive and it may be difficult to prevent cross-linking of the cellulose. Therefore, an alternative approach involving the synthesis of 2-amino-4,6-dichloro-*s*-triazine, its reaction with the cellulose to form a 2-amino-6-chloro-*s*-triazyl-cellulose complex to which the enzyme is coupled, is preferable.

Materials

Ammonia solution (sp.gr. 0.88)
Cyanuric chloride
Dioxane
Toluene
Acetone
Cellulose powder
Na_2CO_3 15% solution
HCl 1.0 mol l^{-1}
HCl concentrated
Phosphate buffer 0.05 mol l^{-1}, pH 7.0
Borate buffer 0.05 mol l^{-1} pH 9.0
Borate buffer 0.05 mol l^{-1}, pH 9.0 containing 1.0 mol l^{-1} NaCl

Method

*1–8 Formation of 2-amino-4,6-dichloro-*s*-triazine*
*9–15 Formation of amino-chloro-*s*-triazyl-cellulose*

16–19 Coupling of papain to activated cellulose

1. Slurry 184 g of cyanuric chloride in 1.2 dm^3 of dioxane : toluene solvent (5 : 1 v : v). Cool to 5 °C.
2. Gas the slurry with a dried stream of gas produced by passing nitrogen through the ammonia solution. Continue the gassing until a thick suspension is formed.
3. Filter the reaction mixture and wash the precipitate with 0.5 dm^3 dioxane. Discard the precipitate and retain both filtrate and washings.
4. Evaporate the filtrate and washings to dryness on a rotary evaporator.
5. Redissolve the 2-amino-4,6-dichloro-*s*-triazine in the minimum volume of 1 : 1 (v : v) acetone : water.
6. Evaporate the acetone under reduced pressure.
7. Redissolve the precipitate formed in boiling water.
8. Cool rapidly to room temperature.
9. Dissolve 1 g of the 2-amino-4,6-dichloro-*s*-triazine in 50 cm^3 1 : 1 acetone : water at 50 °C.
10. Add 10 g (dry weight) cellulose and stir for 5 min.
11. Mix 12 cm^3 of 1.0 mol l^{-1} HCl with 8 cm^3 of the Na_2CO_3 solution, add to the cellulose slurry and stir for a further 5 min.
12. Reduce the pH of the slurry to below pH 7.0 with conc.HCl.
13. Filter the reaction mixture and wash three times with 100 cm^3 aliquots of the acetone : water solvent.
14. Wash the amino-chloro-*s*-triazyl-cellulose copiously with distilled water.
15. Wash with 100 cm^3 of 0.05 mol l^{-1} phosphate buffer pH 7.0 and store at 2 °C.
16. Suspend 5 g of the activated cellulose in 50 cm^3 borate buffer pH 9.0.
17. Dissolve 0.5 g unactivated papain in 50 cm^3 borate buffer pH 9.0.
18. Mix the papain solution and cellulose suspension and stir at 4 °C for 24 hours.
19. Filter the enzyme–cellulose complex and wash with 100 cm^3 borate buffer pH 9.0, followed by 100 cm^3 borate buffer pH 9.0 containing 1.0 mol l^{-1} NaCl and, finally, with 100 cm^3 borate buffer pH 9.0

Kay, G., and Lilly, M. D. (1970) *Biochim. Biophys. Acta*, **198**, 276.

IV FORMATION OF ALKALINE PHOSPHATASE MEMBRANES

Enzyme membranes may be constructed either by adsorbing the enzyme on to a preformed membrane and then cross-linking it in place, or by cross-linking the protein directly to form a sheet of material. Both methods involve the use of a bifunctional reagent such as glutaraldehyde. One method of each type is described.

Sheets of cross-linked protein may be fabricated by cross-linking a mixture of enzyme and inert protein with glutaraldehyde. Inert protein is used to increase the bulk of the membrane. Enzyme membranes based on preformed cellulose acetate or cellulose nitrate membranes have the distinct advantage that they may be fabricated with a defined pore size.

Materials for Protein Sheet Formation

Alkaline phosphatase
Bovine serum albumin solution 60 mg cm^{-3} in 0.02 mol l^{-1} phosphate buffer pH 6.8
Glutaraldehyde solution 2.5%
Glycine solution 0.1 mol l^{-1} adjusted to pH 9.0

Method for Protein Sheet Formation

1. Dissolve 20 mg alkaline phosphatase in 5 cm^3 bovine serum albumin solution.
2. Add 15 cm^3 of glutaraldehyde solution and mix gently.
3. When the solution viscosity begins to increase, spread the solution over a flat glass plate, taking care not to introduce air bubbles into the solution.
4. Leave overnight at 4 °C or until the membrane is formed.
5. Soak the membrane off the glass plate by immersing it in the glycine solution.
6. Wash the membrane thoroughly in distilled water until the washings show no absorbance at 280 nm.
7. Store wet at 4 °C.

Materials for Enzyme Immobilization to Preformed Membranes

Cellulose acetate sheet 30 μm thick
Alkaline phosphatase solution 10 mg cm^{-3} in 0.02 mol l^{-1} phosphate buffer pH 6.8
Glycine solution 0.1 mol l^{-1} adjusted to pH 9.0
Glutaraldehyde solution 25% diluted to 2.5% with 0.02 mol l^{-1} phosphate buffer pH 6.8

Method for Enzyme Immobilization to Preformed Membranes

1. Spread 5 cm^3 alkaline phosphatase solution over a 10 cm^2 piece of cellulose acetate in a petri dish.
2. Leave at 4 °C until desiccated.
3. Spread 5 cm^3 glutaraldehyde solution over the membrane and leave at 4 °C until desiccated.
4. Soak the membrane in glycine solution for 10 hours at 4 °C.
5. Wash the membrane in distilled water until the washings show no absorbance at 280 nm.
6. Store wet at 4 °C.

Thomas, D., Broun, G., and Selegney, E. (1972) *Biochimie*, **54**, 229.

V POLYACRYLAMIDE GEL ENTRAPMENT OF UREASE

The process of polyacrylamide gel entrapment is applicable to a wide range of enzymes. This form of immobilization has the advantage of providing the enzyme with an ionically neutral carrier matrix which is nevertheless hydrophilic. The particular method described here does not show the leakage of protein from the gel usually associated with gel-entrapped enzymes, probably because of the highly cross-linked, floccular nature of the gel matrix. For proper gel formation polymerization must be carried out in the absence of oxygen.

Materials

Urease
Tris buffer 0.1 mol l^{-1} pH 7.0 containing 10^{-4} mol l^{-1} EDTA
Tris buffer 0.1 mol l^{-1} pH 7.0 containing 10^{-4} mol l^{-1} EDTA and 0.5 mol l^{-1} NaCl
Methylene-bis-acrylamide (MBA)
Acrylamide (ACR)
Dimethylamino-propionitrile
Potassium persulphate solution 10 mg cm^{-3}

Method

1. Make up a solution of 110 mg MBA and 10 mg ACR dissolved in 10 cm^3 Tris buffer pH 7.0 and place in a stoppered flask.
2. Dissolve 15 mg urease in the MBA : ACR solution and gas with nitrogen for 20 min.
3. Add 0.2 cm^3 dimethylamino-propionitrile and mix gently.
4. Add 0.5 cm^3 potassium persulphate solution, mix gently, stopper the flask and allow to stand at room temperature for 20 min. An opaque gel will form.
5. Disrupt the gel by shaking the flask and passing the contents through a fine needled syringe.
6. Filter on a Buchner funnel and wash the gel with 50 cm^3 of 0.1 mol l^{-1} Tris buffer pH 7.0 containing 0.5 mol l^{-1} NaCl.
7. Repeat step 6 three times.
8. Resuspend the enzyme gel in 15 cm^3 of 0.1 mol l^{-1} Tris buffer pH 7.0.

Trevan, M. D., and Grover, S. (1979) *Biochem. Soc. Trans.*, **7**, 28.

References and Bibliography

General Reviews

Biotechnological Applications of Proteins and Enzymes. Ed. Bohak, Z., and Sharon, N. Academic Press 1977.

Methods in Enzymology Vol. **XLIV**. Ed. Mosbach, K. Academic Press, 1976.

Katchalski, E., and Silman, I., and Goldman, R. (1971) *Advs. in Enzymology*, **34**, 445.

McLaren, A. D., and Packer, L. (1970) *Advs. in Enzymology*, **33**, 245.

Specific Papers

Axen, R., Porath, J., and Ernback, S. (1967) *Nature*, **214** 1302.

Berezin, I. V., Klibanov, A. M., Samoklin, G. P., and Martinek, K. in *Methods in Enzymology* Vol. **XLIV**, p. 563. Ed. Mosbach, K. Academic Press, 1976.

Bjorek, L. and Rosen, C-G. (1976) *Biotechnol. Bioeng.*, **XVIII**, 1463.

Chibata, I., and Tosa, T. in *Applied Biochemistry and Bioengineering* Vol. **1**. *Immobilised Enzyme Principles*, p. 334. Eds. Wingard, Jr. L. B., Katchaski-Katzir, E., and Goldstein, L. Academic Press, 1976.

Chibata, I., Tosa, T., Sato, T., Mori, T., and Matuo, Y. in *Proc. Int. Ferment. Symp. 4th Fermentation Technology Today*, p. 383. Japanese Society for Fermentation Technology 1972.

David, A., Metayer, M., Thomas, D., and Broun, G. (1974) *J. Membrane Biol.*, **18**, 113.

Engaser, J-M., and Horvath, C. (1974) *Biochim. Biophys. Acta*, **358**, 178.

Fukui, S., Ikeda, S-I., Fujimura, M., Yamada, H., and Kumagai, H. (1975) *Eur. J. Biochem.*, **51**, 155.

Goldman, R., Kedem, O., and Katchalski, E. (1968a) *Biochemistry*, **7**, 4518.

Goldman, R., Kedem, O., Silman, I., Caplan, S., and Katchalski, E. (1968b) *Biochemistry*, **7**, 486.

Goldstein, L. (1964) *Biochemistry*, **3**, 1913.

Goldstein, L. (1972) *Biochemistry*, **11**, 4072.

Habeeb, A. F. S. A. (1967) *Z. Physiol. Chem.*, **297**, 108.

Hervagault, J. F., Jolv, G., and Thomas, D. (1975) *Eur. J. Biochem.*, **51**, 19.

Hornby, W. E. *et al.* (1966) *Biochem. J.*, **98**, 420.

Hornby, W. E., Lilly, M. D., and Crook, E. M. (1968) *Biochem. J.*, **107**, 669.
Inman, D. J., and Dintzis, H. M. (1969) *Biochemistry*, **8**, 4074.
Inman, D. J., and Hornby, W. E. (1972) *Biochem. J.*, **129**, 255.
Jansen, E. F. (1969) *Arch. Biochim. Biophys.*, **129**, 221.
Karube, I., Matsunaga, T., Tsuru, S., and Suzuki, S. (1977) *Biotechnol. Bioeng.*, **XIX**, 1728.
Kay, G., and Crook, E. M. (1967) *Nature*, **216**, 514.
Kay, G., and Lilly, M. D. (1970) *Biochim. Biophys. Acta*, **198**, 276.
Kay, G., Lilly, M. D., Sharp, A. K., and Wilson, R. J. H. (1968) *Nature*, **217**, 642.
Manecke, G., (1962) *Pure Appl. Chem.* **4**, 45p.
Mattiason, B., and Mosbach, K. (1971). *Biochim. Biophys. Acta*, **235**, 253.
Meister, A. (1973) *Science*, **180**, 33.
Mosbach, K., and Dannielson, B. (1974) *Biochim. Biophys. Acta*, **364**, 140.
Mosbach, K., Borgerud, A., and Scott, C. (1975) *Biochim. Biophys. Acta*, **403**, 256.
Naparstek, A., Romette, J. L., Kernevez, J. P., and Thomas, D. (1974) *Nature*, **249**, 490.
Naparstek, A., Thomas, D., and Caplan, S. R. (1973). *Biochim. Biophys. Acta*, **323**, 643.
Ogata, T. (1968) *Biochim. Biophys. Acta*, **159**, 403.
Ohnishi Ts., and Ohnishi To. (1963) *Nature*, **197**, 184.
Porath, J. in *Methods in Enzymology* Vol. **XXXIV**, p. 13. Ed. Jackoby, W. B., Porath, J., and Wilchek, M. Academic Press, 1974.
Porath, J., Aspberg, K., Drevin, H., and Axen, R. (1973) *J. Chromatogr.*, **86**, 53.
Richards, F. M. (1964) *Proc. Natl. Acad. Sci. (U.S.)*, **52**, 833.
Rimmon, A., Gutman, M., and Rimmon, S. (1963) *Biochim. Biophys. Acta*, **73**, 301.
Srere, P. A., Mattiason, B., and Mosbach, K. (1973) *Proc. Natl. Acad. Sci. (U.S.)*, **70**, 2534.
Storelli, C., Vogeli, H., and Semenza, G. (1972) *FEBS Letts.*, **24**, 287.
Thomas, D., Broun, G., and Selegney, E. (1972) *Biochimie*, **54**, 229.
Tosa, T., Mori, T., Fuse, N., and Chibata, I. (1967) *Enzymologica*, **32**, 153.
Trevan, M. D., and Grover, S. (1979) *Biochem. Soc. Trans.*, **7**, 28.
Vieth, W. R., Ventkatasubramanian, K., Constantinides, A., and Davidson, B, in *Applied Biochemistry and Bioengineering* Vol **1** *Immobilised Enzyme Principles*, p. 240. Ed. Wingard, Jr., L. B., Katchalski-Katzir, E., and Goldstein, L. Academic Press, 1976.
Weetal, H. H. (1969) *Science*, **166**, 615.
Wharton, C. W., Crook, E. M., and Brockelhurst, K. (1968) *Eur. J. Biochem.*, **6**, 572.

Index

DATE DUE